J. Knoll   The Brain and Its Self

Joseph Knoll

The Brain and Its Self

A Neurochemical Concept of the Innate and Acquired Drives

With 24 Figures and 14 Tables

 Springer

Prof. Joseph Knoll, M.D.
Nagyvárad tér 4
1089 Budapest, Hungary

Library of Congress Control Number: 2004116871

ISBN-10 3-540-23969-3 Springer Berlin Heidelberg New York
ISBN-13 978-3-540-23969-7 Springer Berlin Heidelberg New York

This work is subject to copyright. All rights are reserved, whether the whole or part of the material is concerned, specifically the rights of translation, reprinting, reuse of illustrations, recitation, broadcasting, reproduction on microfilm or in any other way, and storage in data banks. Duplication of this publication or parts thereof is permitted only under the provisions of the German Copyright Law of September 9, 1965, in its current version, and permission for use must always be obtained from Springer-Verlag. Violations are liable for prosecution under the German Copyright Law.

Springer is a part of Springer Science+Business Media
springeronline.com

© Springer-Verlag Berlin Heidelberg 2005
Printed in Germany

The use of general descriptive names, registered names, trademarks, etc. in this publication does not imply, even in the absence of a specific statement, that such names are exempt from the relevant protective laws and regulations and therefore free for general use.

Editor: Dr. R. Lange
Desk Editor: S. Dathe
Production and typesetting: LE-TEX Jelonek, Schmidt & Vöckler GbR, Leipzig
Cover Design: Frido Steinen, eStudio Calamar

Printed on acid-free paper   27/3152/YL - 5 4 3 2 1 0

*To my Parents and my Brother,
victims of the Holocaust,
and to all the innocent victims
of the myths-directed human society*

Preface

Today, on 4 November 2004, as I start to write this preface, worldometers.info shows that the current population of the earth is 6,433,096,068; during the last 24 h, 203,063 people have been born, but only 83,498 have died. As usual, CNN pours out the latest news around the clock, convincingly illustrating the simultaneous coexistence of order and chaos in human society.

I think that the human cortex created human society and is responsible for its maintenance and continuous progress. The history of science shows that natural laws are simple and gray, but the phenomena brought into existence by them, however, are immensely complex and colorful. It has been the aim of my life to find a reasonable physiological explanation for the coexistence of order and chaos in human communities. In this study, I summarize the most important findings of my research in support of the ultimate conclusion that the unique ability of the human cortex to acquire drives created human society, a society which is still in the myths-directed trial and error phase of its development and aims to find its reason-directed final equilibrium state.

The Brain/Self problem has always been and will ever remain the Question No. 1 for human beings. The last analysis, a brilliant piece of writing by the late Karl R. Popper and John C. Eccles, was published in 1977. The title of their book, *The Self and Its Brain: An Argument of Interactionism,* discloses that the authors were dualists. Popper, the philosopher, described himself as an agnostic. Eccles, the brain scientist, was a believer in God and the supernatural. Both thought it improbable that the Brain/Self problem can ever be really understood. Both distrusted any simple solutions. Nevertheless, even for someone not in agreement with the basic approach of Popper and Eccles, the intellectual content of the twelve dialogues between the two authors in September 1974 (Popper and Eccles 1977, Part III) is fascinating and exerts an irresistible influence upon the reader.

The title of my monograph, *The Brain and Its Self: A Neurochemical Concept of the Innate and Acquired Drives,* immediately makes clear that I have cast off the dualistic approach. A special analysis of the acquired drive and the discovery of previously unknown brain mechanisms have allowed a reasonable physiological interpretation of the development of social life and of the origin of art and science. This study is an attempt to demonstrate that natural laws are simple even in the Brain/Self relation.

Acknowledgements

It is a great pleasure to first of all mention my indebtedness to Károly Kelemen and Berta Knoll (my wife), who entered my laboratory as students and were my only coworkers in the early 1950s when we learned to instill the glass-cylinder-seeking drive into the cortex of rats. This method allowed us to study the nature of an acquired drive and catalyzed my understanding of the crucial importance of this mechanism in the development of social life on earth.

I also wish to express my gratitude to János Dalló, with whom I worked together for 40 years, and to Ildikó Miklya, who joined me in 1980 and is my chief coworker at present. Their enthusiastic work has been particularly helpful in better understanding the operation of an acquired drive.

I would also like to mention all my other colleagues who have worked with me during the 30 years of my chairmanship of the Department of Pharmacology at the Semmelweis University of Medicine. We have developed new drugs, of which (−)-deprenyl is now the best known, a compound used worldwide. More than a dozen of them are now professors, internationally known scientists. I will jot here down in alphabetical order only the names of those with whom I worked for at least two decades: S. Fürst, K. Gyires, H. Kalász, V. Kecskeméti, K. Magyar, J. Timár,. and E.S. Vizi.

I am indebted to Kunyoshi Fujimoto, President of the Fujimoto Pharmaceutical Company (Osaka, Japan), who made me the offer to work together with his staff to perform my planned structure–activity relationship study, aiming to develop a new enhancer substance, more potent than (−)-deprenyl. The collaboration with the Fujimoto Research Laboratory, led by the excellent chemist, F. Yoneda, resulted in the development of R-(−)-1-(benzofuran-2-yl)-2-propylaminopentane, (−)-BPAP, the to-date most-potent and selective synthetic mesencephalic enhancer substance, and currently the best experimental tool for studying mesencephalic enhancer regulation.

In closing, I wish to express my thanks to my good friends, L. Gyermek (USA), T. Ban (Canada), and E. Varga (USA), for their valuable comments on the manuscript.

Contents

Abbreviations XIII

Introduction 1

1 Innate and Acquired Drives 11
   1.1   Innate Drives in the Service of a Limited Number
        of Indispensable (Vital) Goals 11
   1.2   Acquired Drives in Service of an Unlimited Number
        of Dispensable Goals.. 11
   1.3   The Glass-Cylinder-Seeking Drive in Rats:
        A Model for Studying the Characteristics of the Development
        of an Acquired Drive... 12
   1.4   Specific Activation (Active Focus): The Physiological Basis
        of a Drive 14

2 The Conception that Whatever Humans Achieved Derives
   from the Unrestricted Capacity of Their Brain to Acquire Drives 17

3 Enhancer Regulation: A Neurochemical Approach to the Innate
   and Acquired Drives 25
   3.1   Mesencephalic Enhancer Regulation:
        Natural and Synthetic Mesencephalic Enhancer Substances 25
        3.1.1 Definition of Enhancer Regulation:
              β-Phenylethylamine (PEA) and Tryptamine,
              Endogenous Enhancer Substances 25
        3.1.2 The Role of (−)-Deprenyl in the Recognition
              of the Enhancer Regulation in the Mesencephalic Neurons 27
        3.1.3 (−)-Deprenyl (Selegiline) and R-(−)-1-(benzofuran-2-
              yl)-2-propylaminopentane [(−)-BPAP], Prototypes
              of Synthetic Mesencephalic Enhancer Substances 33
   3.2   Pharmacological Analysis of Mesencephalic Enhancer
        Regulation Using (−)-BPAP as a Specific Experimental Tool 39

		3.2.1	Detection of a Specific and a Nonspecific Form of Enhancer Regulation in the Mesencephalic Neurons. Studies Using Isolated Discrete Rat Brain Regions 39
		3.2.2	Analysis of the Two Forms of Enhancer Regulation on Isolated Brain Cells in Culture 40
	3.3	Considering Enhancer Receptors 47	
	3.4	Cortical Enhancer Regulation: Assumptions About Its Physiological Significance 49	
		3.4.1	Essential Forms of the Modification of Behavior Through Exercise, Training, or Practice 49
		3.4.2	The Concept that Learning Is a Cortical Enhancer Regulation Dependent Function 55
	3.5	Therapeutic Aspects of Synthetic Mesencephalic Enhancer Substances 62	
		3.5.1	The Physiological Mechanisms that Give Reason for the Prophylactic Administration of a Synthetic Mesencephalic Enhancer Substance to Slow Brain Aging 62
		3.5.2	Rationale for Slowing the Age-Related Decline of Mesencephalic Enhancer Regulation by the Daily Administration of a Small Dose of a Synthetic Mesencephalic Enhancer Substance from Sexual Maturity Until Death 81
		3.5.3	Clinical Experiences with (−)-Deprenyl in Depression and in Neurodegenerative Diseases: Further Therapeutic Prospects 88

4 Approaching Old Problems From A New Angle 95
    4.1 A New Interpretation of the Substantial Individual Differences in Behavioral Performances 95
    4.2 A New Interpretation of Forgetting, Remembering, and Boredom 98

5 Theoretical Aspects of the Enhancer Regulation Approach 113
    5.1 Simultaneous Coexistence of Determinants of Order and Chaos in the Human Brain: An Approach to the Origins of Science and Art 113
    5.2 The Timeliness of the Conception of the Enlightenment: *Sapere Ande!* (Dare to Go Independently!) 129

6 Conclusion 141

References 147

Index of Names 163

Index of Subjects 171

Abbreviations

(−)-BPAP	*R*-(−)-1-(benzofuran-2-yl)-2-propylaminopentane
CAR	conditioned avoidance response
CR	conditioned reflex
CS	conditioned stimulus
ECR	extinguishable conditioned reflex
EF	escape failure
ICR	inextinguishable conditioned reflex
IR	intertrial response
MAO	monoamine oxidase
MAO-A	A-type monoamine oxidase
MAO-B	B-type monoamine oxidase
PEA	β-phenylethylamine
(−)-PPAP	(−)-1-phenyl-2-propylaminopentane
US	unconditioned stimulus

Introduction

> Imagination is more important than creation because it lays down the foundation of all creation.
>
> *Albert Einstein*

Where Do We Come From? What Are We? Where Are We Going?
Brain research is the only discipline that can answer these fundamental questions.

*

Viewed from a historical distance, it seems to me that Pavlov's epoch-making discovery on conditioning changed once and for all scientific thinking about the brain and its self. He *experimentally* demonstrated, a hundred years ago, that the well-known psychic experience described/defined with the term *association* objectively refers to the acquisition of a *conditioned reflex,* the establishment of a new temporary connection between cortical neurons. This was the decisive turn that initiated the objective analysis of behavioral performances.

On the other hand, despite the rather widespread reservations about Freud's life-work, his teaching about the mental structure of a human being was obviously primarily responsible for the dissemination of the currently generally accepted views regarding the operation of the human psyche.

In retrospect it can be established as a fact that the tempestuous decades between 1890 and 1960 represent the most productive, revolutionary period in brain research. The pioneering work of Thorndike on animal intelligence, summarized in his two monographs published in 1898 and 1911, and then further extended as the psychology of wants, interests, and attitudes in 1940, opened up new channels of research (Thorndike 1898, 1911, 1940). The studies on animal drives in the 1920s (Moss 1924, Richter 1927), the first deeper analysis of the purposive behavior in animals and in man (Tolman 1932), the methodological innovations that significantly improved the quality of behavioral studies in animals (Skinner 1938), new theories of learning (Hilgard 1948), the concept of Gestalt psychology (Koffka 1935, Köhler 1947), the definitive discovery of the ascending reticular activity system (Moruzzi and Magoun 1949), and last but not least, the golden 6th decade of the 20th century with its memorable series of breakthroughs in the pharmacology of the central nervous system (the deeper penetration into the problem of hallucinogens, the development of MAO inhibitors, phenothiazines, tricyclic antidepressants, uptake inhibitors, etc.) that brought into being the science of neuropsychopharmacology (see Ban et al. 1998; Healy 1996, 1998, 2000, for review) changed in a revolutionary manner the general views about the principles of behavior and radically altered human attitudes toward derangements in psychic function.

Furthermore, during the last four decades, we have been witness to a previously almost unimaginable progress in the technical possibilities of investigating the molecular physiology, pharmacology, and pathology of neurons. Brain research reached its indisputably most successful analytical phase, but with regard to the brain/self problem the previously established theoretical interpretations remained practically unaltered. *The main aim of this study is to analyze the relation between the brain and its self from a theoretically new point of view.*

*

It is a horrifying fact that in Germany millions of single-minded little-men who had previously lived an honest simple life and never belonged to extremist groups dramatically changed within a few years after 1933 and, imbued with the Nazi ideology, became unbelievably cool-headed murders of innocent civilians during the Second World War. This phenomenon has been documented from many angles in dozens of novels, films, and so on. However, we are still waiting for an adequate elucidation of the brain mechanism responsible for this dramatic and rapid change in the behavior of millions.

As a survivor of Auschwitz I had the opportunity to directly experience a few typical representatives of this type of manipulated human beings, and had more than enough time and direct experience to reflect upon the essential changes in the physiological manipulability of the human brain. It was therefore not just by mere chance that when in the early 1950s I finally had the opportunity to approach this problem experimentally, I decided to develop a rat model to follow the changes in the brain in the course of the acquisition of a drive from the start of training until its manifestation. As briefly recounted in Sect. 1.3, we built the glass-cylinder-seeking drive into the brain of our rats (Knoll et al. 1955a,b,c, 1956; Knoll 1956, 1957). The first series of studies, the results of which were summarized in a monograph (Knoll 1969), clarified that a special form of excitation in a proper group of cortical neurons ("active focus") is the basis of an acquired drive. Further studies, the results of which were summarized in a recent review (Knoll 2003), allowed us to conceptualize that two previously unknown brain mechanisms, the regulation of mesencephalic and cortical enhancement [hereafter referred to as "(mesencephalic and cortical) enhancer regulation"], are responsible for the innate and acquired drives. This study is an attempt to prove that only the behavior of species capable of acquiring drives can be manipulated. But, since the human brain is unmatched in its unlimited capacity to acquire drives, the manipulability of humans is also unparalleled. This singular capacity of the human brain has resulted in the establishment of the unique human society.

*

This study is also responsible for developing the conceptualization that vertebrates can be divided into three groups according to the mode of operation of their brain: (a) those that operate with innate drives only (the majority); (b) those with an ability to acquire drives (a minority); and (c) the "group of one" that operates almost exclusively with acquired drives (*Homo sapiens sapiens*).

With the evolution of brains capable of acquiring drives species appeared whose members could manipulate each other's behavior and act in concert. This was the condition *sine qua non* for the evolution of social living, a form of life that enabled the species to surpass qualitatively the performance of any given individual. It goes without saying that training members in the skills needed to act in concert improved the quality of life. The learned behavior, for example, of five to six hungry female lions act in unison to separate from the herd the animal chosen to be brought down, significantly increases the chance of capturing the prey. It was the evolution of a brain with the ability to acquire drives that made the appearance of life on earth so immensely variable.

With the development of the human brain, a functional network with over 100 billion interrelated nerve cells and 10^{10} bit capacity arose. With this system, whose operation is inseparably connected to conscious perception, life on earth reached its most sophisticated form. Furthermore, the human being, who is primarily a social creature, is a building block in the creation of a gigantic product: human society. The function and capacity of society obviously exceeds the sum of the activity of its members. Based on the practically inexhaustible capacity of the human brain to acquire drives, human society represents a qualitatively new, higher form of life. For example, a country, presently the most sophisticated form of a human community, consists of millions or even over a billion humans and operates *de facto* as a huge living complex interacting with other similar entities, about 200 at present.

The birth and development of human society, a moment insignificant and fleeting in the endless history of the universe, necessarily means everything to us. It can be taken for granted that at the birth of human society about 80,000 years ago (with all probability somewhere in South-Africa), very small groups formed a micro-community, working together. About 70,000 years had to pass until, owing to the cortical activity of humans, due to learning, practice, and experience, their community life became more and more efficient, and the accumulation of basic knowledge opened the way for a more rapid development, truly reflected in population growth.

In the last phase of the Stone Age, about 8,000–9,000 years before our age, marked by the domestication of animals, development of agriculture, and the manufacture of pottery and textile, the human population on earth approached the *one million* level. Thereafter, however, the population increase necessarily went from strength to strength. By the beginning of the Common Era it had reached the *300 million* level, grew to *1.6 billion* by 1900, and is at present around *6.5 billion*. There is a compelling reason to curb this rate of population increase, as it already endangers human survival.

*

The main function of human society, the regulation of the production and distribution of goods that determine the essential conditions of its members, has always been and still remains far from being equally satisfactory for all. Globally, about one billion people are at present poorly housed, of whom 100 million live on the streets. To keep the system running "optimally", it is an absolute necessity to properly manipulate the masses. In due time society builds into each brain the proper drives that compel one to voluntarily accept, or at least submissively tolerate, actual living conditions. The élite, possessing executive power, develops the hierarchical organization of the society necessary for keeping law and order. Human life reflects the ceaseless adaptation to social requirements. Success of the individual as a member of the society depends upon the zealous, assiduous acquisition of the proper drives expected and accepted by the micro- and macro-community to which the individual belongs.

Possessing the same stock of cortical neurons, every healthy human brain is born with a potential to produce whatever a human brain has produced in the past and will produce in the future. This immense capacity remains practically unexploited. The main reason for the extremely low utilization of inborn brain capacities lies in the historically developed, systematic manipulation of the brain that ensures that society continues going in a predetermined direction. It was for this same reason that the admonition of the Enlightenment, *Sapere ande*! (Dare to go independently!), necessarily failed to develop among the masses.

The dissemination of knowledge about the operation of the human brain, the realization how its ability to acquire drives makes the individual so vulnerable to outer influences, may in the long run radically decrease and finally eliminate the responsiveness of individuals to historically developed and still effective manipulative techniques. The latter would otherwise increase the conscious exploitation of inborn abilities. In the best case, this trend may some day help humankind to approach the mythical golden age, the evolution of an optimum form of social living.

The main message of this monograph is that the appearance of the mammalian brain with its ability to acquire drives ensured the development of social life and ultimately led to the evolution of the human society. This most sophis-

ticated form of organized life on earth is still in the trial-and-error phase of its development. It seeks to outgrow the myth-based era of its history and arrive at its final state, a rationally organized human society.

*

For the human being the definition of the self was always and remains forever the question of questions. Past efforts to resolve this problem crystallized into two opposing opinions: (a) the self has its brain, (b) the brain has its self.

Believers in the view that the self has its brain contend that the soul is the principle of human life and held the spiritual part of man to be as distinct from its physical counterpart. Furthermore, they consider feeling, thought and the action of man as entities distinct and separate from the body.

Believers in the view that the brain has its self deny this duality. They declare that the psychic experience, the cognitive/volitional and affective state of consciousness, the most amazing product of nature, is inseparable from measurable changes in brain neurons. The objective and subjective aspects of brain activity are thought to be as interrelated as the outside and the inside of one and the same thing. This study is an exercise in support of this opinion on the basis of the following *new* argument:

In the mammalian brain capable of acquiring drives, untrained cortical neurons (Group 1) possess the potentiality to change their functional state in response to practice, training, or experience in three consecutive stages, namely by getting involved in (a) an extinguishable conditioned reflex (ECR) (Group 2), (b) an inextinguishable conditioned reflex (ICR) (Group 3), or (c) an acquired drive (Group 4). The activity of the cortical neurons belonging to Groups 3 and 4 is inseparable from conscious perception. In any moment of life self is the sum of those cortical neurons that have already changed their functional significance and belong to Groups 3 or 4.

According to the cortical enhancer regulation approach, whenever we observe in the behavioral performance of an individual the signs of an efficiently operating acquired drive, the objective change in the brain is that a population of cortical neurons reach the maximum level of excitability, due to the permanent production of their specific enhancer substance in optimal concentration. Many past analysts of drive-induced behavioral performances already came to the conclusion that cortical neurons must be able to stay in a permanent state of high excitability. Uchtomsky's concept of "dominant focus" (1945) and Lorente de No's proposal of "reverberating circuits" (1935) are examples of this line of reasoning.

It also seems reasonable to assume that whenever we observe the efficient operation of an acquired drive, the objective changes in the cortical neurons are inseparable from the subjective side of the special state of high excitability, the imagination of the goal to be reached. Terms such as "process of imagination" (Beritov 1932), "insight" (Koffka 1935), and so on have been adopted by virtue of an analogy between the behavior of animals and humans. These terms, just like the term "drive," are only useful descriptions, and Pavlov was right to assert that "No knowledge has come from description alone" (Pavlov 1955).

It is obvious that we can analyze only the objective changes in the cortical neurons belonging to Groups 3 and 4, and can only describe the subjective side of the story, the conscious perception. The latter is inseparable from the objectively analyzable changes in the neurons and does not exist without the operation of the cortical enhancer regulation beyond a critical threshold. We should never forget that the objectively analyzable changes in the cortical neurons belonging to Groups 1 and 2 can never be described since they are never consciously perceived. Even the activity of the cortical neurons belonging to Groups 3 and 4 remains unconscious until the operation of the cortical enhancer regulation crosses a critical threshold inseparable from conscious perception and therefore becomes describable.

Due to a never-ending stimulation of the brain via the outer and inner world the transformation of cortical neurons from a functionally lower category into a higher one continues uninterruptedly from birth until death. While the potential to acquire drives is highly restricted in the brains of animals, this capacity is unlimited in the human cortex. This qualitative leap makes the *Homo sapiens sapiens* unique among living beings.

*

Popper is correct in noting: "The productive scientist as a rule starts from a *problem*. He will try to understand the problem" (Popper and Eccles 1977, p. 39). In this quotation the designation "productive" refers implicitly to a scientist who tries to shed light on a significant, unsolved problem, a process which usually demands a lifelong passionate devotion to research. Prior to the acquisition of such devotion the scientist attained knowledge via readings, observations, and/or performed preliminary experiments that put him into the position to define a problem. *This was the decisive training period* that ultimately culminated in the acquisition of the proper drive needed to pursue the creative process with unflagging enthusiasm to its finish.

What definitely happens in the brain is that a population of cortical neurons ascend, as a result of the decisive training period, in the hierarchy and assume the role of Group 4. Objectively, this group of neurons acquires the ability to produce its specific enhancer substance in optimal concentration and stay, when needed, permanently in a state of the highest attainable level of excitability. This specific state of neuronal activity, the "active focus" (see Knoll 1969, for review), is inseparable from the subjective, psychic experience which can be described as a "problem," "concept," "insight," "process of imagination," and so on. Whenever we talk about an acquired-drive-controlled intellectual activity, we necessarily approach it from the subjective side, as this is the way we experience it. In this sense, it is true that problem-solving of any significance requires from its beginning a concept formation that provides the foundation for any specific experiments.

In a similar sense, it is also true that it is the unique faculty of the human brain to form an image about something what is not signaled through the senses, and this is obviously the mother of creation. The experiments performed and the concept that directs them mutually interact and change each other until the goal is reached. But, whatever the theoretically prepared mind discovers depends on the sum of the ICRs and the nature of the acquired drives already fixed in the brain. Even in serendipity, previously accumulated knowledge is the condition *sine qua non* for realizing the value of the unsought after finding.

In my case the decisive period of experimentation that built into my brain the urge to clarify *the unknown neurochemical mechanism of the acquisition of a drive* fell between 1951 and 1953. After 50 years of continuous analysis of this problem, it has become my firm belief that the mammalian brain reached its highest level of organization with the evolution of specific cortical enhancer regulation enabling it to acquire drives. I also propose that this development culminated in the appearance of the *Homo sapiens*, the only mammalian species whose life is primarily based on the acquisition of "unnatural" drives. It seems to me that just as the discovery of the force of mutual attraction among all bodies led to a sound interpretation of the world around us, the discovery of the force of mutual attraction between cortical neurons will lead to a sound interpretation of a brain function which is inseparable from conscious perception.

*

In behavioral studies "drive" is the commonly used technical term to define the force that activates the mammalian organism. It is the inner urge that initiates a response, incites activity, and that represents a basic or instinctive need, such as the hunger drive, the sexual drive, and so on.

The neurochemical basis of both categories of drives – (a) the innate ones necessary for the survival of the individual and the species, and (b) the acquired ones for attaining an unlimited number of dispensable goals – is unknown. The mesencephalic mechanism that keeps the innate drives in action is presumably the less complicated part of the problem. The real crux of the issue seems to be the cortical mechanism that renders the acquisition of an unnatural urge possible.

Being familiar with a technical term we may occasionally have the erroneous impression of possessing full knowledge of the subject it connotes. For example: An eagle pounces upon a quiet, eating rabbit with lightning speed. The rabbit has a split second to run for its life. Common sense, practical explanation independent of specialized knowledge, is simple. Hunger drives the eagle and fear drives the rabbit. In reality, drive is just a useful description for the still unknown brain mechanism that activates the organism and keeps it in motion until the goal is reached.

Based on previous efforts to reveal the underlying mechanism of innate and acquired drives (see Knoll 1969, 2003, for review), this study attempts to translate "drive" into the language of neurochemistry.

*

For living beings with highly refined brain organization the cortex has absolute priority in maintaining the sophisticated integration between an apparently confusing network of cells, synchronizing them into a lucidly arranged, harmoniously operating system. For a highly refined organism life means the operation of the integrative work of the brain, and natural death means the cessation of this function. This is clearly shown by the fact that cells of vital organs, including the brain, maintain vigorous activity for a short while even beyond the termination of the integrative work of the brain.

Mesencephalic enhancer regulation, primarily in the catecholaminergic neurons, keeps the telencephalon active and thus the system alive. The operation of the catecholaminergic system is comparable to an engine ignited once and for all in an early phase of development, and is signaled by the appearance of an EEG. Due to its enhancer regulation, the catecholaminergic system dynamically changes the activation of the cortex during lifetime according to need. Life is terminated because of the progressive decay of the efficiency of the catecholaminergic system during the postdevelopmental lifespan until at some point, in an emergency situation, the integration of the parts in the highly sophisticated entity can no longer be maintained. Thus natural death, signaled by the disappearance of an EEG, sets in (see Knoll 1994, for review).

The catecholaminergic tone determines the three basic modes of brain activity. The system performs: (a) at its lowest possible level in the "nonvigilant resting state" (sleeping); (b) at a steady low level in the "vigilant resting state" (leisure); and (c) operates, according to the need, at a dynamically enhanced activity level in the "active state" (exemplified by "fight or flight" or goal-seeking behavior).

Experimental evidence and theoretical considerations in this monograph led to the conceptualization that an until recently unknown brain mechanism, the enhancer regulation in the mesencephalic neurons, is primarily responsible for the innate drives, and a special form of it in the cortex is primarily responsible for the acquired drives. Furthermore, data support the conclusion that age-related changes in the enhancer regulation of the catecholaminergic brain engine are primarily responsible for: (a) the youthful power of mammals from weaning until sexual maturity; (b) the transition from the uphill period of life to postdevelopmental longevity; (c) the progressive decay of behavioral performances during the downhill period; and (d) the transition from life to death.

Finally, the data reinforce the proposal (see Knoll, 2001, for review) that prophylactic administration of a synthetic mesencephalic enhancer substance during postdevelopmental life could significantly slow the unavoidable decay of behavioral performances, prolong life, and prevent or delay the onset of age-related neurodegenerative diseases, such as Parkinson's and Alzheimer's.

1 Innate and Acquired Drives

Urges that keep the mammalian organisms working as highly sophisticated, goal-oriented entities can be divided into two main groups: innate and acquired drives.

1.1
Innate Drives in the Service of a Limited Number of Indispensable (Vital) Goals

Innate drives divide into two subgroups:

A. Drives that ensure the survival of the individual. They are:
   1. The urge to maintain internal stability (homeostasis)
   2. The urge to keep away or to get rid of anything that is endangering or unpleasing
   3. The urge to obtain water and food, and
B. Drives that ensure the survival of the species. They are:
   1. The urge to copulate
   2. The urge to nurture offspring

The analysis of innate-drive-dependent functions (maintenance of homeostasis, fight for survival, feeding, sexuality, progeny-care, etc.) constitute the main body of literature on behavioral physiology and endocrinology. Though innate drives are primarily based on mesencephalic regulations, none of the goals can be reached without the participation of the cortical neurons. Exclusively innate drives keep the majority of the mammalian species alive.

1.2
Acquired Drives in Service of an Unlimited Number of Dispensable Goals

The ability to acquire an irrepressible urge for a goal that is not necessary for survival of the individual or species represents the most sophisticated function of the telencephalon. Though the development of an acquired drive always originates in one way or another from an innate drive, this relation later becomes unrecognizable.

Humans are the only living beings on earth whose life is predominantly based on acquired drives. To a certain extent, a minority of the mammalian species (the monkey, dog, horse, dolphin, rat, etc.) possesses this endowment, which, under natural conditions, remains unexploited. Nevertheless, humans obviously discovered thousands of years ago, probably through a kind of serendipity, that the behavior of such animals can be modified by proper training, and this started the development of the domestication of various species.

The ambition to be in a permanent state of activity is a natural endowment of the human brain, which acquires drives with utmost ease. The acquisition of proper drives in the most sensitive developmental period of life, from weaning until sexual maturity, will thereafter be determinant for the lifelong basic activity of the individual.

However, since the fate of most individuals is still governed by the position in society into which they are born, only a minority is lucky enough to acquire professional drives in full harmony with natural endowments. The majority, as a matter of fact, forms – under coercion – the work-related drives that will ensure the place of the individual in the society. Conformity between one's innate abilities and acquired work-related drives is of key importance for lifelong equilibrium. However, not only the desire to be permanently active is a natural endowment of the human brain, but there is also a need for a new challenge to one's drives in due time. Even the most satisfying professional drive becomes boring after its permanent, continuous use and there is a need to continue to keep the brain in a satisfyingly active state. Inexhaustible forms of supplementary activities serve this aim.

Absolute dominance of a fully satisfying professional drive and the acquisition of well-chosen supplementary drives are the conditions for a harmonious, well balanced life.

Lack of full satisfaction in one's acquired professional and supplementary drives generates an urge to flee from frustration and seek salvation in "Ersatz": smoking, alcohol, drugs, and so on.

1.3
The Glass-Cylinder-Seeking Drive in Rats:
A Model for Studying the Characteristics of the Development of an Acquired Drive

Some strains of laboratory rats lend themselves particularly well to the analysis of the formation of an acquired drive and the study of the basic role of an innate drive in the acquisition of an urge for an unnatural goal.

1.3 The Glass-Cylinder-Seeking Drive in Rats

In the early 1950s we developed a method to build a special acquired urge, the "glass-cylinder-seeking drive," into the brain of rats (Knoll et al. 1955a,b,c, 1956; Knoll 1956, 1957; see also Knoll 1969, for review). Based on an unconditioned avoidance reflex (escape from a hot plate) and using the sound of a shrill bell to play the role of a high-priority conditioned stimulus, rats were trained to search for and jump to the rim of a 30-cm-high glass-cylinder. The cylinder was open at bottom and top with diameters of 16 cm and 12 cm, respectively, and with a side opening through which a rat (up to 350–400 g body weight) could manage to get inside the cylinder.

In the training procedure the rat was pushed through the side opening of the glass-cylinder standing on a metal plate heated to 60 °C, and the jumping reflex was elicited for a couple of weeks three times daily on 10–50 occasions at 10 s intervals with bell and heat stimulation. After a short training period a chain of ICRs developed and the rat displayed the jumping reflex without heat stimulation 100 times in succession (Knoll et al. 1955a,b,c). This was a transient stage leading to the manifestation of the glass-cylinder-seeking drive (for review see Chap. 4 in Knoll 1969).

Those rats that performed best acquired the glass-cylinder-seeking drive in a stable manner, thereafter maintaining this unnatural urge for a lifetime. These rats showed the same high-grade adaptability and readiness in overcoming different obstacles during goal-attainment as the ones influenced by innate drives, such as hunger or sexual desire (Knoll et al. 1956; Knoll 1956, 1957).

In the most efficiently trained, best performing rats the acquired drive was so powerful that it suppressed even innate drives. When such a rat has been deprived of food for 48 h and then food was offered within the usual setup that contained the glass-cylinder, the rat looked for the glass cylinder and left the food untouched. Similarly, when a receptive female was offered to a sexually fully active glass-cylinder-seeking male rat in the usual setup, the male looked for the glass-cylinder and neglected the receptive female.

Urged by the acquired drive, glass-cylinder-seeking rats built with the same extreme rapidity chains of ECRs consisting of hundreds of newly acquired associations as rats driven by hunger. By changing the position of the glass cylinder in the setup, the chain of ECRs extinguished as rapidly as in the case of innate drives, and the rat immediately built a new chain of ECRs according to need (Knoll 1956, 1957). This function elucidates the real physiological role of the ECRs as tools that allow proper accommodation to a rapidly changing outside world (see Knoll 1969, for review).

It is obvious that the ability of the cortical neurons to acquire a drive is an innate potentiality only. With the aid of our training method we manipulated the brain of the rat, recruiting in a proper group of cortical neurons that functional change which kept the animal active until the goal was reached. This modification of behavior could never have occurred during the lifetime of the rat under natural conditions.

1.4
Specific Activation (Active Focus): The Physiological Basis of a Drive

We described earlier the essence of both the innate and acquired drives as a state of specific activation (active focus) in a special population of subcortical and cortical neurons, respectively (see Fig. 11 in Knoll 1969). In the light of the enhancer regulation concept, we may characterize the active focus as an endogenous enhancer substance-induced enhanced excitability in a circumscribed population of mesencephalic and telencephalic neurons that persists until the goal has been reached.

In the case of innate drives enhancer regulation in the mesencephalon is responsible for *both* the formation of the subcortical active focus that maintains the enhanced orienting-searching reflex activity until the goal is reached *and* cortical active focus ("the cortical representation of the drive"). As natural conditions are always changing, even the goals determined by innate drives can be reached only with the participation of cortical neurons. The successful operation of an innate drive requires, namely, the continuous acquisition of proper chains of ECRs.

In the case of the acquired drives, however, the essence of the complicated behavioral performance is the activation of enhancer regulation in a special population of cortical neurons, forming a cortical active focus that keeps the organism in action until the goal is reached. Whatever the cortically determined goal is, it cannot be reached without a strong orienting-searching performance resulting from properly enhanced mesencephalic activity. Accordingly, an acquired drive brings the mesencephalic system into the same state of enhanced activity as innate drives do.

In this context the catecholaminergic machinery in the mesencephalic system deserves special consideration. Independent of any goal-directed behavioral performance, with all its cognitive and emotional consequences, the continuous perception of the outer and inner world and the maintenance of homeostasis keep the brain active by themselves. Thus the "engine of the brain" is incessantly in motion. Catecholamines, which influence – activate or inhibit – billions of neurons, are continuously released in the mesencephalon. It was well-documented in the 1970s that an extreme paucity or even lack of synaptic junctions is a common feature of noradrenergic (Ajika and Hökfelt 1973; Calas et al. 1974; Descarries et al. 1977), dopaminergic (Hökfelt 1968; Tennyson et al. 1974), and serotonergic (Richards et al. 1973; Chan-Palay 1975; Calas et al. 1976; Descarries and Leger 1978) nerve terminals. This peculiar situation means that the amount of catecholamines released within a given time interval will determine the extent of the catecholaminergic influence on the whole brain. Enhancer regulation, capable of changing dynamically the amount of free catecholamines in the brain according to need, plays a determinant role in survival.

1.4 Specific Activation (Active Focus): The Physiological Basis of a Drive

It is reasonable to assume that whenever a drive is operating the mesencephalic enhancer regulation works on a higher activity level. This means that during the operation of a drive a significantly greater amount of catecholamines and serotonin can be detected in the mesencephalon. To test the validity of this working hypothesis, we compared the amounts of monoamines released by isolated, discrete brain regions in sated vs food-deprived rats (Miklya et al. 2003b).

We first measured – in a special open field – the orienting-searching reflex activity of rats deprived of food for 48 and 72 h, respectively, and isolated thereafter the discrete brain areas from the mesencephalon and measured the amount of norepinephrine, dopamine, and serotonin released by the tissue samples into an organ bath. *The orienting-searching reflex activity of the rats **increased** proportionally to the time elapsed from the last feed. Simultaneously the amount of 1. dopamine released from the striatum, substantia nigra, and tuberculum olfactorium, 2. norepinephrine released from the locus coeruleus and 3. serotonin released from the raphe also **increased** in the hungry rats proportionally to the time of fasting.* For example, the amount of dopamine released from the substantia nigra of sated increased after fasting for 48 and 72 h from 4.62 ± 0.20 nM/g wet weight to 5.95 ± 0.37 ($P < 0.05$) and 10.67 ± 0.44 ($P < 0.01$), respectively. For details see Miklya et al. (2003b).

2 The Conception that Whatever Humans Achieved Derives from the Unrestricted Capacity of Their Brain to Acquire Drives

It was already discovered in ancient times that the behavior of some mammalian species can be manipulated by proper training. The animals acquire a drive for an unnatural goal and humans make use of it. The horse and the dog are probably the best examples of domesticated species that for thousands of years played an important role in the everyday life of humans. Their faithfulness and devotion to their master, their cleverness and special skill to be helpful in complicated situations is legendary. The essence of domestication is clear by now. The manipulation of the brain of domesticated animals, which enables their exploitation after proper training, is based on the ability of their cortex to acquire drives. Yet the overwhelming majority of vertebrates is devoid of this ability.

In the light of our present knowledge I conjecture that living organisms on earth arrived at the attainable peak of sophistication with the evolution of a cortex capable of acquiring drives. This concept already took shape in the early 1950s when *we successfully built into the brain of rats the glass-cylinder-seeking drive, while we were unable to force mice to acquire a similar drive.* This was a clear hint that the potentiality to acquire drives represents a higher level of cortical organization. A lifelong, thorough examination of rats that had acquired the glass-cylinder-seeking drive played a decisive role in the experimental foundation of this idea. It catalyzed new lines of research, ultimately leading to the realization that *enhancer regulation* is the neurochemical basis of the drives (Knoll 2003).

Observation of any act in the endless fight for existence drama in nature illustrates the crucial importance of enhancer regulation. When the eagle pounces upon a rabbit with lightning speed, life or death is in the balance. Both the attacker and the potential victim have only a split second to become properly activated. The chance for the eagle to obtain its food and for the rabbit to save its life lies in the efficiency of a specific mechanism that activates the brain and through it the whole organism. The participant with the more efficiently activated brain will reach its goal (see Knoll 1969, 1994, 2001, 2003, for review). The rabbit's only chance of survival depends upon its split-second transformation from a relaxed state to a state of excitation that enables it to mobilize all resources and run for its life. Enhancer regulation is the brain mechanism responsible for this change. The term "drive", used to cherish illusions that we know what is happening, is just a resounding phrase to describe the phenomenon.

We can easily understand the essential, drive-induced, behavioral consequences by observing the movement of rats in an open field with numbered squares in which activity is measured by the number of squares crossed in a 30-min period and by the total area covered during this period. Due to the innate orienting-searching reflex activity, a naive rat put in this open field looks searchingly around for a short while and then stops moving.

A rat urged on by an innate or an acquired drive behaves *qualitatively* differently. The orienting-searching reflex activity seems to be inextinguishable. The reason for the difference, in my interpretation, is as follows: In the mesencephalon of the naive rat put in the open field only the best performing catecholaminergic neurons react to the new surrounding. In the "drive" situation endogenous enhancer substances raise the excitability of the catecholaminergic neurons and a significantly higher number of neurons respond to the same stimulation. This is the essence of the change.

The characteristic behavioral consequences of the operation of an innate or acquired drive is shown in rats in an open field in Table 1. The data were taken from a recently published series of experiments (Miklya et al. 2003b) that corroborated our earlier findings published between 1955 and 1957 (see Knoll 1969, for review).

Table 2.1 demonstrates that rats supplied *ad libitum* with food and water crossed on average **11** squares within 30 min and covered **9.2%** of the total area (Series no. 1). Rats driven by the deprivation of food for 72 h, however, crossed **73** squares and covered **54.6%** of the total area (Series no. 3), and rats that acquired the "glass-cylinder-seeking drive" and were activated through the specific CS, crossed **265** squares and covered **88.4%** of the total area (Series no. 8). It is obvious that the drive-induced enhanced orienting-searching-reflex activity is essential for ultimately reaching, by trial and error, a goal that is not perceivable by the senses at the start of the experiment.

This is further clearly shown by comparing the drive-induced purposeful increase in activity with the purposeless hypermotility caused by amphetamine treatment. Amphetamine induces a continuous, irresistible release of catecholamines from their intraneuronal stores in the mesencephalon and this leads to aimless hyperactivity. Groups of rats treated with 1, 2 or 5 mg/kg amphetamine crossed in average **98**, **216**, and **369** squares, respectively, but the total area covered by the animals was on average **8.0**, **9.8**, and **7.1%** (Series no. 4, 5, 6), the same as covered by rats devoid of a drive (Series no. 1). Moreover, amphetamine completely inhibited the goal-directed activity enhancement, induced by an innate (Series no. 7), or an acquired drive (Series no. 9).

*

Table 2.1. Demonstration of the drive-induced, essential behavioral consequences in an open field in rats: The qualitative difference between an innate-, or acquired-drive-induced purposeful hypermotility, due to enhanced orienting-searching reflex activity,[a] on the one hand, and amphetamine-induced purposeless hypermotility, due to continuous release of catecholamines from their intraneuronal stores, on the other

No.	Type of experiment	Average number of squares crossed in the open field within 30 min	Average percentage of the total area of the open field covered within 30 min
1	Control	11	9.2
2	Food-deprivation for 48 h[b]	57	36.8
3	Food-deprivation for 72 h[b]	73	54.6
4	Amphetamine (1 mg/kg)[d]	98	8.0
5	Amphetamine (2 mg/kg)[d]	216	9.8
6	Amphetamine (5 mg/kg)[d]	369	7.1
7	Amphetamine (5 mg/kg) treatment of rats deprived of food for 72 h[d]	311	8.2
8	Glass-cylinder-seeking rats[c]	265	88.4
9	Amphetamine (5 mg/kg) treatment of glass-cylinder-seeking rats[c,d]	377	14.4

[a] The orienting-searching reflex activity was measured in an open field in rats according to Knoll (1957).
[b] Food-deprivation experiments were performed on three-month-old male Wistar rats ($n = 20$).
[c] Glass-cylinder-seeking rats ($n = 5$) were trained according to the method of Knoll (1969).
[d] Amphetamine was injected subcutaneously 30 min prior to the trial period.

At the end of the 3rd week after birth enhancer regulation in the rat's mesencephalon starts working on a significantly higher activity level. This is the discontinuation of breast feeding, the crucially important first step to living separately from the mother (Knoll and Miklya 1995). Weaning is obviously the onset of the developmental (uphill) phase of the individual life of the mammalian organism (Knoll 1994, 2001). The period, characterized by a higher basic activity, lasts until the rat develops full-scale sexual maturity (Knoll et al. 2000).

One of the telltale signs which makes the operation of the mesencephalic enhancer mechanism evident is the enhanced basic activity of the catecholaminergic and serotonergic systems, as measured by the significantly enhanced release of catecholamines and serotonin from discrete brain regions isolated from the brain of rats after weaning. Reaching sexual maturity this change disappears and the basic activity of the catecholaminergic and serotoninergic systems returns to the preweaning level (Knoll and Miklya 1995).

Sexual hormones seem to be responsible for the transition from the developmental, uphill phase of life into the postdevelopmental, downhill period, characterized by the slow age-related decay of brain performance terminated by natural death (Fig. 6 in Knoll 2001). Weighty arguments speak in favor of the assumption that the slow, continuous age-related decline of enhancer regulation in the mesencephalic neurons plays a key role in the progressive decay of behavioral performances with the passing of time (see Sect. 3.5.1.1 for details).

According to our present knowledge the nigrostriatal dopaminergic neurons that maintain the enhanced orienting-searching reflex activity indispensable for successful goal-seeking behavior are the most rapidly aging units in the human brain. Over age 45 the dopamine content of the human caudate nucleus decreases steeply, at a rate of 13% per decade. If dopamine sinks below 30% of the normal level, symptoms of Parkinson's disease appear. About 0.1% of the population over 40 years of age develops Parkinson's disease and prevalence increases sharply with age. Parkinson's disease is an especially convincing example of an age-related neurodegenerative disease due to the unusually fast deterioration of an enhancer-sensitive group of midbrain neurons (see Sect. 3.5.1.2.1 for details).

Although the decay of enhancer regulation starts with the full scale development of sexual hormonal regulation (Knoll et al. 2000), this does not mean that the sexually mature individual is immediately converted to a significantly lower performer in his or her fight for existence. Learning, the modification of behavior through experience, training, or practice ensures a rapid and successful goal-directed performance without the need for the high-level specific activation of enhancer-sensitive midbrain neurons. This is nature's most ingenious method to enhance the chances for survival even in the downhill period of living. The experienced organism works in an economic manner and is always reaching its goal with much lower energy investment than the inexperienced one. Nevertheless, the progressive age-related decay of enhancer regulation is irresistibly weakening the ability to acquire new information. As a consequence, vitally important adaptability to a new situation is necessarily on a progressive decline. Thus, even the most experienced, aged organism becomes more and more vulnerable in its struggle for life with the passing of time.

*

The progressive deterioration of the brain engine's performance with the passing of time is paralleled by the proportional decay in the ability of naive cortical neurons to change by training their functional state. This is manifested by the progressive age-related decline of the capacity to build ECRs, fix ICRs and acquire drives. Thus, the stability of already fixed information is in inverse relationship to the time of its acquisition. This helps explain why aged people vividly relive their youth but forget their most recent experiences.

To ecphorize already fixed chains of ICRs and activate acquired drives does not require significant effort. In contrast, the acquisition of new information, the transformation of a naive cortical neuron into an experienced one, is the most complicated operation. It requires the highest energy investment and is therefore the most vulnerable function of the cortical neurons. Clinical experiences, based on thousands of reliable case histories, support this conclusion.

The famous pathography that describes the illness of the great musician, Maurice Ravel, published by his physician Alajouanine (1948), is a convincing example. This case clearly demonstrates the high vulnerability of the mechanism in the cortical neurons responsible for the fixation of chains of ICRs and for the acquisition of drives. Ravel, who died in 1937, wrote his two last compositions (two brilliant piano concertos) in 1929–1930 because, suffering a severe cerebral lesion on both sides of his brain, he later completely lost his ability to compose or even learn a piece of music unknown to him. This in no way hindered him, however, from playing his own compositions, and he easily ecphorized other music known to him before his illness. This case demonstrates that it is easy to revive an engram that was already irreversibly fixed in the past and bring it into consciousness even from a seriously lesioned brain, but to transform naive neurons into trained ones is the most complicated, most vulnerable process in the brain.

The leading idea of this monograph is the view that in a species capable of acquiring drives the function of the cortical neurons changes in response to training in three consecutive phases. Each of these represents a characteristic form of behavior, as follows:

- Untrained $\Rightarrow$ ECR $\Rightarrow$ ICR $\Rightarrow$ acquired drive

Only a limited number of species possess the ability to fix ICRs and acquire drives. Among them the horse, elephant and camel were trained by humans to help them in their hard work, the dog and cat are the most popular pets, while the anthropoid apes, the dolphin and the laboratory rat seem to be especially suitable species for the experimental analysis of fixation of ICRs and the acquisition of drives. We used the rat for this purpose.

*

The history of human society provides eloquent testimony for a highly significant improvement in its quality of life with the passing of time. It is obvious that this development is due to the organized effort of society. This collective performance is qualitatively different from that of its members. It is easy to realize that the efficient operation of any human community is based on the cooperation of brains that can be manipulated. Human society is unique because the brain work of its members is unique. The human brain acquires drives with a facility that makes the *Homo sapiens* qualitatively different from other species.

Though the performance of society is qualitatively different from the performance of its members, development originates from the discoveries and innovations of individuals. Any significant change is due to the achievement of a single talent who succeeded to discover something that was developed and exploited thereafter by the community. The essence of the history of any society is the history of discoveries and inventions handed down from generation to generation. In the immense literature describing the behavior of animals we find thousands and thousands of exactly characterized acquired drives that furnish convincing evidence for the role of discoveries and innovations that changed the habits of a community.

To give an example from the animal kingdom: The chimpanzee is the most clever of animals. Its ability to develop sophisticated techniques is beyond the reach of other animals. There are colonies of chimpanzees that eat termites as an especially nourishing delicacy. Termites live in special earth dwellings full of thin passages that lead to the surface. The termite-eating chimpanzee looks for a twig of a proper size to introduce into one of the passages and then waits until an adequate number of termites stick to the twig. The chimpanzee also acquires by training the special skill, requiring a high degree of competence, to draw out the twig without losing the termites that cling to it. It is obvious that a chimpanzee once made the chance discovery that termites are delicious, was talented enough to acquire the drive aiming to obtain termites, and developed the necessary technique to reach the goal (invention). Members of the colony thereafter adopted the method and trained their descendants. Colonies of chimpanzees that have failed to discover this possibility miss a nourishing delicacy.

The development of the ancient human community was driven similarly by unexpected, brilliant discoveries and ingenious inventions made by talented individuals. Domestication of animals is a good illustration of this. Somebody realized by chance that the behavior of a wild animal can be manipulated, and this discovery initiated the long process that ultimately led to the breeding of properly domesticable strains.

According to the present scientific view, based on careful genetic studies on European, Asian, North-American, and African dogs, all the presently living strains of dogs are descendants of the she-wolf that lived about 15,000 years ago in what is now China. It is obvious that an unknown talent discovered by chance that the brain of a wolf can be properly manipulated, thereby initiating the long breeding process that created the dog, the first domesticated animal in history. There can be little doubt that the domestication of the wild horse about 6,000 years ago and the domestication of the wild cat about 3,000 years ago had an essentially similar history.

Traditional medicine dating back to the dawn of human civilization witnessed another main line of discoveries that improved significantly the quality of life of human communities. A couple of those ancient discoveries laid the foundation of modern pharmacology as it developed thousands of years later.

Animals capable of acquiring drives, especially the anthropoid apes, possess, to a limited extent, the ability to perceive the unknown and the skill to create something new. Nevertheless, even the human-intervention-based acquisition process that induced the most sophisticated learned behavioral performance of a chimpanzee shows little evidence of incorporating images or concepts into the organization of the animal's behavior. In contrast, imagination is the dominant brain mechanism that determines human activity. As a consequence the human brain is unique in (a) approaching objective reality also in an abstract form, and (b) using symbols as tools of a language for oral or written interpersonal communication. It seems reasonable to assume that this basic difference is rooted in the human brain's remarkable ease in building ICRs and acquiring drives. The sheer quantity of stored experience found in a human brain is therefore immense in comparison even to the amount of information stored in the brain of a chimpanzee. Thus, the translation of quantity into quality is what explains why it is only in the human brain that the conditions are available for the recombination of previous experiences into new images in support of problem-solving.

It is the mode of existence of human society that whatever is discovered/invented in science/technology, art, and social life will sooner or later be evaluated, mastered, copied/modified/multiplied by others. History is the witness that this is the basic mechanism responsible for development. Just for illustration: Emil Fischer (recipient of the 1902 Nobel Prize for chemistry), the pioneer of chemical syntheses that opened the way for the development of carbohydrate and protein chemistry, irreversibly changed the history of a branch of science. Similarly, the van Eyck brothers, who first intentionally used the oil technique, irreversibly changed the history of painting, and so on and so forth.

The mechanism whereby innovations are rapidly copied/modified/multiplied is at work even when they are harmful for human society. The history of modern terrorism, which grew out of enormous political extremism, furnishes a cautionary tale. It was discovered that due to the manipulability of the human brain suicide killers – the cheap/nasty, but very efficient human bomb – can be prepared, then used as a means of political pressure. The developed world felt compelled to wage war against this dreadful method. This is an ongoing problem. In the beginning the group(s) responsible for a terror attack could usually be precisely identified. What we see now is that lines between organizations are blurring, with old groups disbanding and re-emerging under new names. It is a further example of the eternal fate of inventions.

On the other hand, as soon as a useful innovation develops to fitness, it rapidly becomes public property and is soon perfected. For example, in 1946, when the technique of broadcasting images via radio waves to receivers which then project them onto a picture tube became ripe for general use, 6,000 television sets soon enriched households in the United States with this technical innovation. By the end of 1950 the number of sets increased to 4.4 million and reached the 50-million level within a further decade.

3 Enhancer Regulation: A Neurochemical Approach to the Innate and Acquired Drives

3.1
Mesencephalic Enhancer Regulation: Natural and Synthetic Mesencephalic Enhancer Substances

3.1.1
Definition of Enhancer Regulation: β-Phenylethylamine (PEA) and Tryptamine, Endogenous Enhancer Substances

We can define enhancer regulation as: the existence of enhancer-sensitive neurons capable of changing their excitability in a split second and working on a higher activity level, due to natural enhancer substances. Of the agents with such effect, for the time being, only β-phenylethylamine (PEA) and tryptamine have been experimentally analyzed (Knoll 2001, 2003).

Though enhancer sensitive neurons exist also outside the mesencephalon, we used the term "mesencephalic enhancer regulation" to emphasize the key importance of the dopaminergetic neurons, the most rapidly aging neurons of the brain, primarily responsible for the progressive age related decline of behavioral performances.

The catecholaminergic and serotonergic neurons in the mesencephalon are excellent models to study the enhancer regulation since their physiological function is to supply the brain continuously with the proper amounts of monoamines that influence – activate or inhibit – billions of neurons. The significant enhancement of the nerve-stimulation-induced release of [^{3}H]-norepinephrine, [^{3}H]-dopamine, and [^{3}H]-serotonin from the isolated brain stem of the rat in the presence of PEA (Fig. 3.1) or tryptamine (Fig. 3.2) is shown to illustrate the response of enhancer-sensitive neurons to endogenous enhancer substances.

From a freshly isolated brain stem of a properly pretreated rat a low amount of the labeled transmitters is released for a couple of hours (see Knoll and Miklya 1995, for methodology). Neurons respond to stimulation in an "all or none" manner. The calculated average amount of each of the transmitters released from the non-stimulated brain stem is the product of the spontaneous firing of the most excitable, most responsive group of neurons of the surviving population with large individual variation in excitability. The overwhelming

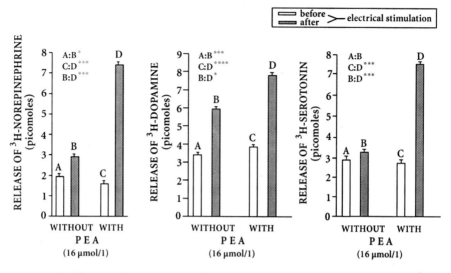

Fig. 3.1. Significant enhancement of the nerve-stimulation-induced release of [³H]-norepinephrine, [³H]-dopamine, and [³H]-serotonin, respectively, from the isolated brain stem of the rat in the presence of β-phenylethylamine (PEA) ($n = 8$). Each *graph bar* represents the amount of the labeled transmitter in picomoles released in a 3-min collection period. See Knoll et al. (1996c) for methodology. *Vertical lines above the graph bars* show the SEM (standard error of the mean). Paired Student's *t*-test was used for statistical analysis. *$P < 0.05$, **$P < 0.02$, ***$P < 0.01$, ****$P < 0.001$

majority of the neurons remain silent. Electrical stimulation excites a further group of neurons as shown by the significant increase of the outflow of transmitters. Natural enhancer substances increase specifically the excitability of the enhancer-sensitive neurons. Accordingly, Figs. 3.1 and 3.2 demonstrate that in the presence of PEA or tryptamine the amount of transmitters released to electrical stimulation increased dramatically.

The data in Figs. 3.1 and 3.2 show a remarkable quantitative difference between PEA and tryptamine in their effectiveness on serotonergic neurons. A lower concentration of tryptamine (1.3 µmol/l) proved to be much more potent in enhancing the stimulation-evoked release of serotonin than a much higher concentration of PEA (16 µmol/l). This indicates that, on a molecular level, the enhancer regulation in the catecholaminergic and serotonergic neurons are not identical.

Enhancer regulation in the brain heralds a new line of research. It brings a different perspective to brain-organized, goal-oriented behavior since it seems to represent the device in the mammalian brain that operates as the *vis vitalis*.

PEA and tryptamine, the first examples of physiological enhancer substances, represent just the peak of an iceberg. The development of a tryptamine-derived synthetic enhancer substance that increased the performance of cul-

3.1 Mesencephalic Enhancer Regulation

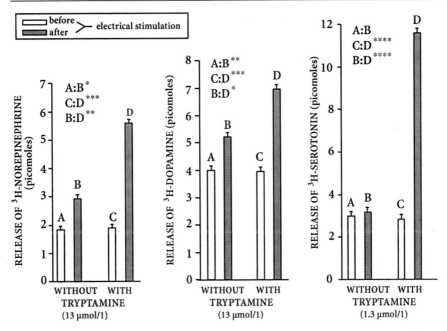

Fig. 3.2. Significant enhancement of the nerve-stimulation-induced release of [^{3}H]-norepinephrine, [^{3}H]-dopamine, and [^{3}H]-serotonin, respectively, from the isolated brain stem of the rat in the presence of tryptamine. ($n = 8$). Each *graph bar* represents the amount of the labeled transmitter in picomoles released in a 3-min collection period. See Knoll et al. (1996c) for methodology. *Vertical lines above the graph bars* show the SEM. Paired Student's *t*-test. *$P < 0.05$, **$P < 0.02$, ***$P < 0.01$, ****$P < 0.001$

tured hippocampal neurons with a peak effect at 10^{-14} M concentration (see Fig. 5 in Knoll et al. 1999) foreshadows the existence of much more potent physiological enhancer substances in the brain than PEA and tryptamine and incites research in this direction.

3.1.2
The Role of (−)-Deprenyl in the Recognition of the Enhancer Regulation in the Mesencephalic Neurons

A thorough, 30-year analysis of the mechanism of action of (−)-deprenyl, a drug presently registered world-wide for the treatment of Parkinson's disease, ultimately resulted in the recognition of enhancer regulation in the mesencephalic catecholaminergic neurons (see Knoll 1998, for review). The history how this crucially important physiological mechanism remained undetected for decades gives a good example of concealed traps in research.

We developed (−)-deprenyl in the early 1960s (see Knoll 1983, for a review of the first two decades of its history). When we started to develop (−)-deprenyl,

MAO inhibitors were at the center of interest. Both as experimental tools and as therapeutic agents MAO inhibitors had an important influence on the development of the widely accepted hypothesis: that 1. depression is associated with diminished monoaminergic tone in the brain, and 2. depressed patients treated with antidepressants become elated because of enhanced biological activity of monoamine transmitters in the central nervous system.

The discovery of the mood-elevating effect of MAO inhibitors was a classic example of serendipity in drug research. In 1951, isoniazid and its isopropyl derivative, iproniazid, were successfully introduced for the treatment of tuberculosis. In contrast to isoniazid, iproniazid was found to produce undesirable stimulation in some patients. In 1952, Zeller and his co-workers demonstrated that iproniazid was capable of inhibiting MAO, whereas isoniazid was ineffective (Zeller and Barsky 1952; Zeller et al. 1952). In 1956, Crane analyzed the psychiatric side-effects of iproniazid and came to the conclusion that it might be beneficial in the treatment of depression (Crane 1956). In 1957 Kline introduced it as a "psychic energizer" (Kline 1958). At the same time Kuhn discovered the antidepressant effect of imipramine (Kuhn 1957). This opened the way to the most powerful antidepressant therapy to date.

At the beginning there was a keen interest in the MAO inhibitors, of which a substantial number were developed and introduced into clinical practice, but because of serious side-effects there was a rapid turnover in the introduction and withdrawal of these drugs. In 1963, a calamitous number of clinical reports, demonstrating the occurrence of dangerous hypertensive attacks in patients treated with MAO inhibitors were published. Blackwell suggested that the hypertensive crises are associated with the ingestion of high amounts of tyramine in cheese, the metabolism of which is inhibited by the MAO inhibitors ("cheese effect") (Blackwell 1963). This conclusion was correct. Cheese and many other foods containing tyramine were found to be able to provoke hypertensive episodes in patients treated with MAO inhibitors. The "cheese effect" restricted the clinical use of this group of drugs.

Deprenyl (we used the racemic compound under the code name E-250 in the first series of experiments) proved to be a compound with a peculiar pharmacological spectrum. We described it in our first paper as a new spectrum psychic energizer (Knoll et al. 1965). I selected this compound for further development because I was fascinated by the finding that in contrast to MAO inhibitors, which potentiated the blood pressure increasing effect of amphetamine, a releaser of norepinephrine from their stores in the noradrenergic nerve terminals, E-250 *inhibited* it (see Fig. 1 in Knoll et al. 1965). Based on this observation we analyzed this peculiar behavior in more detail. As I expected, the studies revealed that *deprenyl, in contrast to the known MAO inhibitors, did not potentiate the effect of tyramine but inhibited it.* This effect of deprenyl was first demonstrated in a study performed on cats and on the isolated vas deferens of rats. The hope was expressed in this paper that this peculiar tyramine-inhibiting property of a potent MAO inhibitor might be of special therapeutic value (Knoll et al. 1968).

In the same year that the description of the unique behavior of (−)-deprenyl was published, Johnston described a substance, later named clorgyline, that came into world-wide use as an experimental tool in MAO research. Johnston realized that clorgyline preferentially inhibits the deamination of serotonin. He proposed the existence of two forms of MAO, "type A" and "type B," the former being selectively inhibited by clorgyline and the latter relatively insensitive to it. Johnston's nomenclature has become widely accepted and is still in use (Johnston 1968).

For further studies a selective inhibitor of MAO-B was needed. We were lucky to discover in 1970 that (−)-deprenyl was the missing, highly selective inhibitor of MAO-B (Knoll and Magyar 1972). The compound was used thereafter as the specific experimental tool to analyze MAO-B. Our first paper that described this novel property became a citation classic 10 years later (Knoll J, This Week's Citation Classic, January 15, 1982). For several years the selective MAO-B inhibitory effect was at the center of our interest. It delayed the discovery of the drug's enhancer effect. It was the MAO inhibitory effect of the compound that led to the first clinical application of (−)-deprenyl.

In light of the serious side effects of levodopa in Parkinson's disease, Birkmayer and Hornykiewicz tried to achieve a levodopa-sparing effect by the concurrent administration of levodopa with an MAO inhibitor. As such combinations frequently elicited hypertensive attacks, they were soon compelled to terminate this line of clinical research (Birkmayer and Hornykiewicz 1962).

We had already shown in animal experiments that (−)-deprenyl is a unique MAO inhibitor which does not potentiate the catecholamine-releasing effect of indirectly acting amines, but instead inhibits it. We proposed to use it as an MAO inhibitor free of the cheese effect (Knoll et al. 1968). The validity of this proposal was shown in man by Sandler and his co-workers (Elsworth et al. 1978; Sandler et al. 1978). Considering the peculiar pharmacological profile of (−)-deprenyl, Birkmayer et al. (1977) dared to combine this MAO inhibitor with levodopa in Parkinson's disease. This trial was successful. The levodopa-sparing effect was achieved in parkinsonians without signs of significant hypertensive reactions. This study initiated the world-wide use of (−)-deprenyl in Parkinson's disease.

Today the most evaluated effect of the drug is its ability to slow the rate of the functional deterioration of the nigrostriatal dopaminergic neurons in patients with early, untreated Parkinson's disease, and thus to slow the progress of the disease. The indication for using (−)-deprenyl in *de novo* parkinsonians was established in the DATATOP study in the USA (Tetrud and Langston 1989; Parkinson Study Group 1989, 1993) and was corroborated in important multicenter studies (Allain et al. 1991; Myttyla et al. 1992; Larsen et al. 1999).

Age-related deterioration of the striatal machinery is a continuum and any precisely determined short segment of it is sufficient to measure the rate of decline in the presence or absence of (−)-deprenyl. As a matter of fact, in the DATATOP multicenter study of the Parkinson Study Group a segment of this continuum, the time elapsing from diagnosis of Parkinson's disease until levodopa was needed, was properly measured in untreated patients with Parkinson's disease and the effect of (−)-deprenyl versus placebo was compared (Parkinson Study Group 1989). Tetrud and Langston (1989) were the first to publish the finding that (−)-deprenyl delays the need for levodopa therapy. In their study, the average time that elapsed before levodopa was needed was 312.1 days for patients in the placebo group and 548.9 days for patients in the (−)-deprenyl group. This was clear proof that (−)-deprenyl, which enhances the activity of the surviving dopaminergic neurons, kept these neurons on a higher activity level for a longer duration of time.

The design of the DATATOP study was unintentionally the same that we had used in our rat experiments with (−)-deprenyl since 1980. We tested the sexual activity of male rats as a quantitatively measurable, rapidly aging dopaminergic function, and compared the effect of (−)-deprenyl versus saline treatment on the age-related decline of copulatory activity in rats. We demonstrated that (−)-deprenyl treatment significantly slowed the age-related decay of sexual performance (Knoll 1982) and later went on to show that this effect of (−)-deprenyl was unrelated to the inhibition of MAO-B. We performed a structure–activity relationship study aiming to select a derivative of (−)-deprenyl that is free of any MAO-B inhibitory property (Knoll et al. 1992a). In (−)-deprenyl the propargyl group binds covalently to the flavin group of MAO-B, and this leads to the irreversible inhibition of the enzyme activity. (−)-PPAP, the new (−)-deprenyl analogue selected, differed from its mother compound by containing a propyl group instead of a propargyl group. As expected, this compound enhanced dopaminergic activity in the brain like (−)-deprenyl, but did not change the activity of MAO-B. One can follow the progress in clarifying the mechanism of action of (−)-deprenyl responsible for enhanced dopaminergic activity in a series of reviews (Knoll 1978, 1983, 1987, 1992, 1995, 1998, 2001, 2003).

By now it is clear that if we select a quantitatively measurable dopaminergic function and determine its age-related decline by fixing an exact end, a shift of this end stage in time in (−)-deprenyl-treated rats shows the dopaminergic activity enhancer effect of the drug. For example, male rats ultimately lose their ability to ejaculate due to the physiological aging of the striatal dopaminergic system. We found that saline-treated rats reached this stage at the age of 112 ± 9 weeks, whereas their (−)-deprenyl-treated peers lost the ability to ejaculate only at the age of 150 ± 12 weeks ($P < 0.001$) (Knoll 1992).

The design of the DATATOP study was essentially the same. The authors knew that after having diagnosed Parkinson's disease the next step would be

3.1 Mesencephalic Enhancer Regulation

the need for levodopa, and they measured the (−)-deprenyl-induced delay in reaching this stage.

The authors of the DATATOP study expected (−)-deprenyl to be efficient in their trial because of its MAO-B inhibitory effect. Their hypothesis was that the activity of MAO and the formation of free radicals predispose patients to nigral degeneration and contribute to the emergence and progression of Parkinson's disease. In accord with their working hypothesis they expected that (−)-deprenyl, the MAO inhibitor, α-tocopherol, the antioxidant, and the combination of the two compounds will slow the clinical progression of the disease because MAO activity and the formation of oxygen radicals contribute to the pathogenesis of nigral degeneration. They selected patients with early, untreated Parkinson's disease and measured the delay of the onset of disability necessitating levodopa therapy.

In the first part of the trial 401 subjects were assigned to α-tocopherol or placebo and 399 subjects were assigned to (−)-deprenyl, alone or with α-tocopherol. Only 97 subjects who received (−)-deprenyl reached the "end" of the trial (i.e., the onset of disability necessitating levodopa therapy) during an average 12 months of follow-up compared with 176 subjects who did not receive (−)-deprenyl. The risk of reaching the end of the trial was reduced by 57% for the subject who received (−)-deprenyl, and these patients also had a significant reduction in their risk of having to give up full-time employment (Parkinson Study Group 1989). Following the course of changes, the authors concluded in their next paper (Parkinson Study Group 1993) that (−)-deprenyl, but not α-tocopherol, delayed the onset of disability associated with early, otherwise untreated Parkinson's disease. But as time passed, the DATATOP study also revealed that (−)-deprenyl did not reduce the occurrence of subsequent levodopa-associated adverse effects in the patients (Parkinson Study Group 1996). A comparison of the enhancer effect of α-tocopherol with that of (−)-deprenyl showed that α-tocopherol did not change the impulse-evoked release of norepinephrine, dopamine and serotonin in the brain, thus it is devoid of an enhancer effect (Miklya et al. 2003a).

Although Tetrud and Langston and other authors of the DATATOP study were not aware of the dopaminergic activity enhancer effect of (−)-deprenyl, their trial was the first to give convincing evidence that (−)-deprenyl, in harmony with our findings in rats, keeps the nigrostriatal dopaminergic neurons on a higher activity level in humans. In addition, this effect of (−)-deprenyl had already been detected in a selected human population with the lowest striatal dopaminergic activity. The highly significant effect of (−)-deprenyl and the ineffectiveness of α-tocopherol during the first years of the DATATOP study were clear proof that (−)-deprenyl acted by enhancing the activity of the nigrostriatal dopaminergic neurons. The patients selected for the study with early, untreated Parkinson's disease were ideal for demonstrating this effect. The subjects still had a sufficient number of dopaminergic neurons whose activity could be enhanced by (−)-deprenyl; thus, the need for levodopa therapy was delayed. α-Tocopherol, devoid of a dopaminergic activity enhancer effect, remained ineffective. As Parkinson's disease is incurable, drug effects are necessarily transient in nature. It is obvious that parallel with further decay of the striatal dopaminergic system, the responsiveness of the patients toward (−)-deprenyl decreased with the passing of time (Parkinson Study Group 1996).

With the development of (−)-1-phenyl-2-propylaminopentane, (−)-PPAP, the (−)-deprenyl analogue free of the MAO-B inhibitory potency, we already furnished direct evidence that the enhanced dopaminergic activity following administration of (−)-deprenyl was unrelated to the inhibition of MAO-B. Because (−)-PPAP, like (−)-deprenyl, inhibited the uptake of tyramine in isolated smooth muscle tests, we first assumed that the drug-induced enhanced dopaminergic activity was due to an uptake inhibitory effect. Further studies revealed that this interpretation was false.

The availability of HPLC to measure catecholamines and serotonin in physiological quantities allowed a new approach. The thorough analysis of the dose-dependent effect of (−)-deprenyl on the release of catecholamines and serotonin from isolated, discrete, rat brain regions (dopamine from the striatum, substantia nigra and tuberculum olfactorium; norepinephrine from the locus coeruleus; and serotonin from the raphe) pointed to enhancer regulation in the mesencephalic neurons. We treated rats with 0.01, 0.025, 0.05, 0.1 and 0.25 mg/kg (−)-deprenyl, respectively, once daily for 21 days, isolated the discrete rat brain regions 24 h after the last injection and measured the biogenic amines released during a 20-min period from the freshly isolated tissue samples. The amount of dopamine released from the substantia nigra and tuberculum olfactorium clarified that the dopaminergic neurons worked on a significantly higher activity level even in rats treated with the lowest dose of (−)-deprenyl, 0.01 mg/kg. As this small dose of (−)-deprenyl leaves MAO-B activity and the uptake of amines practically unchanged, this study was the first unequivocal demonstration of the operation of a hitherto unknown enhancer mechanism in dopaminergic neurons stimulated by (−)-deprenyl in very low doses (Knoll and Miklya 1994).

Further studies clarified the operation of mesencephalic enhancer regulation (Knoll and Miklya 1995; Knoll et al. 1996a,b,c). We realized that PEA, the parent compound of (−)-deprenyl, is primarily an endogenous mesencephalic enhancer substance. Since PEA, in higher concentrations, is a highly effective releaser of catecholamines from their intraneuronal stores, this effect covered up completely the enhancer effect of this endogenous amine, which was classified as the prototype of the indirectly acting sympathomimetics.

Amphetamine and methamphetamine, PEA derivatives with a long lasting effect, share with their parent compound the releasing property. (−)-Deprenyl was the first PEA/methamphetamine derivative that *maintained the enhancer*

3.1 Mesencephalic Enhancer Regulation

effect of its parent compounds but *lost completely the releasing property*. This peculiar change in the pharmacological spectrum of this PEA derivative ultimately enabled the discovery of the enhancer regulation in the mesencephalic neurons, since the enhancer effect of (−)-deprenyl was not covered up by the release of catecholamines from their intraneuronal stores.

In the light of our present knowledge clinicians were mistaken from the very beginning who used (−)-deprenyl in the belief that the therapeutic benefits observed in patients treated with this drug were due to the selective inhibition of MAO-B in the brain. The overwhelming majority of the clinical benefits are due to the enhancer effect of (−)-deprenyl (see Knoll 1998, for review).

3.1.3
(−)-Deprenyl (Selegiline) and R-(−)-1-(benzofuran-2-yl)-2-propylaminopentane [(−)-BPAP], Prototypes of Synthetic Mesencephalic Enhancer Substances

3.1.3.1
(−)-Deprenyl, the PEA-Derived Enhancer Substance

(−)-Deprenyl (Selegiline), developed in the early 1960s as a new spectrum psychostimulant and potent MAO inhibitor, later proved to be, as the first selective inhibitor of MAO-B, indispensable for investigating the nature and function of B-type MAO. Hundreds of clinical studies with the drug were designed thereafter in the firm belief that selective blockade of MAO-B was responsible

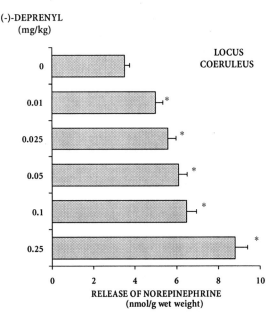

Fig. 3.3. Significant enhancement of norepinephrine release from the locus coeruleus of rats isolated 30 min after the subcutaneous administration of a single dose of (−)-deprenyl. The amount of norepinephrine released from the tissue within 20 min following the administration of different doses of (−)-deprenyl was measured according to Knoll and Miklya (1995). *Horizontal lines to the right of the graph bars* show the SEM. Paired Student's t-test. $*P < 0.01$

for all the effects that followed (−)-deprenyl medication. Realizing, however, that PEA is an endogenous mesencephalic enhancer substance *and* a releaser of catecholamines, while (−)-deprenyl is a PEA-derived synthetic mesencephalic enhancer substance devoid of any catecholamine-releasing property, it became clear that the enhancer effect of (−)-deprenyl was responsible for the majority of the beneficial effects of the drug described in various experimental and clinical studies (see Knoll 1998, 2001, for review).

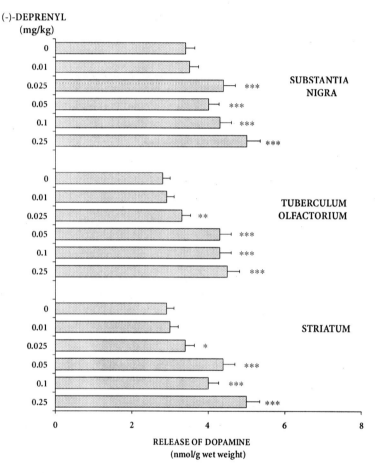

Fig. 3.4. Significant enhancement of dopamine release from the substantia nigra, tuberculum olfactorium, and striatum of rats, respectively, isolated 30 min after the subcutaneous administration of a single dose of (−)-deprenyl. The amount of dopamine released from the tissue within 20 min following the administration of different doses of (−)-deprenyl was measured according to Knoll and Miklya (1995). *Horizontal lines to the right of the graph bars* show the SEM. Paired Student's *t*-test. $*P < 0.05$, $**P < 0.02$, $***P < 0.01$

3.1 Mesencephalic Enhancer Regulation

PEA, rapidly metabolized by MAO, is short acting and its enhancer effect can be detected in *in vitro* experiments only (see Fig. 3.1). Since (−)-deprenyl is not rapidly metabolized, its effect can be measured quantitatively *in vivo*. The subcutaneous administration of (−)-deprenyl enhanced the activity of the catecholaminergic neurons in a dose-dependent manner. This effect is shown on noradrenergic neurons (Fig. 3.3) and on dopaminergic neurons (Fig. 3.4). (−)-Deprenyl treatment, however, did not enhance the activity of serotonergic neurons (Fig. 3.5). (−)-Deprenyl is a PEA-derived enhancer substance and its *in vivo* ineffectiveness on serotonergic neurons is in accord with the finding that PEA was much less potent than tryptamine in enhancing the activity of the serotonergic neurons in the *in vitro* experiments, too (cf. Fig. 3.1 with Fig. 3.2).

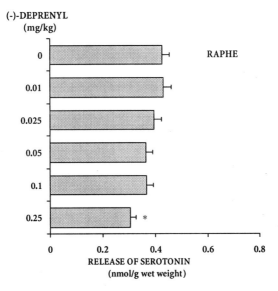

Fig. 3.5. Lack of enhancement of serotonin release from the raphe of rats isolated 30 min after the subcutaneous administration of a single dose of (−)-deprenyl. The amount of serotonin released from the tissue within 20 min following the administration of different doses of (−)-deprenyl was measured according to Knoll and Miklya (1995). *Horizontal lines to the right of the graph bars* show the SEM. Paired Student's *t*-test was used for statistical analysis. None of the applied doses of (−)-deprenyl enhanced the release of serotonin significantly, the highest dose even decreased the release significantly. $*P < 0.05$

Since (−)-deprenyl is a highly potent and selective inhibitor of MAO-B, we performed a structure-activity relationship study to develop a deprenyl-derived enhancer substance that is free of the MAO-B inhibitory property (Knoll et al. 1992a). (−)-1-Phenyl-2-propylaminopentane [(−)-PPAP] has been chosen as our reference substance with this pharmacological profile.

Figure 3.6 shows the chemical structure and pharmacological spectrum of PEA and its four most representative synthetic derivatives.

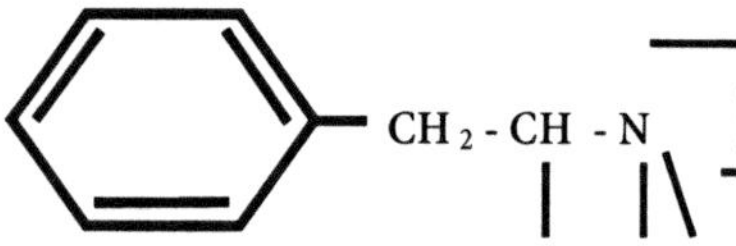

				ENHANCER EFFECT	RELEASING EFFECT	RELATION TO MAO
β-PHENYLETHYLAMINE (PEA)	H	H	H	+	+	SUBSTRATE TO MAO-B
AMPHETAMINE	CH_3	H	H	+	+	WEAK MAO INHIBITOR
METHAMPHETAMINE	CH_3	CH_3	H	+	+	WEAK MAO INHIBITOR
(-)-1-PHENYL-2-METHYL-N-METHYL-PROPARGYL-AMINE, (-)-DEPRENYL	CH_3	CH_3	C_3H_3	+	0	POTENT MAO-B INHIBITOR
(-)-1-PHENYL-2-PROPYL-AMINOPENTANE, (-)-PPAP	C_3H_7	H	C_3H_7	+	0	0

Fig. 3.6. The chemical structure and pharmacological spectrum of PEA and its most representative synthetic derivatives. Taken from Knoll 2001

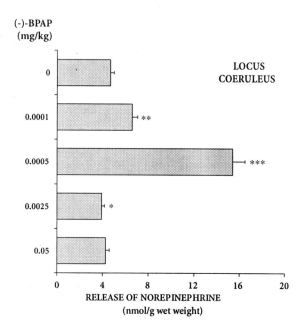

Fig. 3.7. Significant enhancement of norepinephrine release from the locus coeruleus of rats isolated 30 min after the subcutaneous administration of a single dose of (−)-BPAP. The amount of norepinephrine released from the tissue within 20 min following the administration of different doses of (−)-BPAP was measured according to Knoll and Miklya (1995). *Horizontal lines to the right of the graph bars show SEM. Paired Student's t-test.* $*P < 0.05$, $**P < 0.01$, $***P < 0.001$

3.1.3.2
(−)-BPAP, the Tryptamine-Derived Enhancer Substance

The discovery that tryptamine is also an endogenous enhancer substance (Knoll 1994) opened the way for a structure-activity relationship study aiming to synthesize a new family of enhancer compounds structurally unrelated to PEA and the amphetamines. R-(−)-1-(benzofuran-2-yl)-2-propylaminopentane [(−)-BPAP] was selected as a tryptamine-derived synthetic mesencephalic

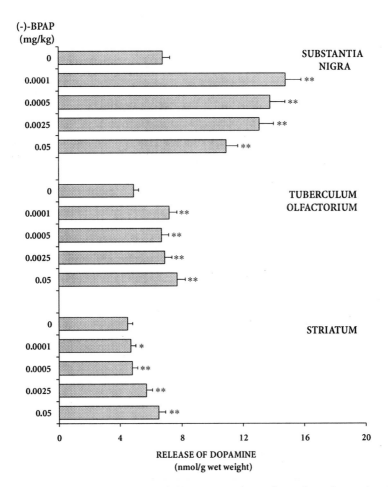

Fig. 3.8. Significant enhancement of dopamine release from the substantia nigra, tuberculum olfactorium, and striatum of rats, respectively, isolated 30 min after the subcutaneous administration of a single dose of (−)-BPAP. The amount of dopamine released from the tissue within 20 min following the administration of different doses of (−)-BPAP was measured according to Knoll and Miklya (1995). *Horizontal lines to the right of the graph bars* show SEM. Paired Student's t-test. $*P < 0.05$, $**P < 0.01$

enhancer compound for further studies (Knoll et al. 1999). For details of its chemistry see: Oka et al. (2001) and Yoneda et al. (2001).

The *in vivo* dose-dependent enhancer effect of (−)-BPAP is illustrated on noradrenergic (Fig. 3.7), dopaminergic (Fig. 3.8), and serotonergic neurons (Fig. 3.9), respectively. A comparison of the enhancer effect of (−)-BPAP and (−)-deprenyl shows

1. The substantially higher potency of (−)-BPAP than (−)-deprenyl in enhancing the activity of catecholaminergic neurons
2. The characteristic dose-dependency of the enhancer effect of (−)-BPAP on noradrenergic (Fig. 3.7) and serotonergic neurons (Fig. 3.9), and
3. The highly potent *in vivo* enhancer effect of (−)-BPAP on serotonergic neurons (Fig. 3.9) and the lack of this effect on the part of (−)-deprenyl (Fig. 3.5)

In a study the effect of uptake inhibitors (desmethylimipramine, fluoxetine), a selective MAO-A inhibitor (clorgyline), a selective MAO-B inhibitor (lazabemide), and dopamine receptor stimulants (pergolide, bromocriptine) – in comparison to the effect of (−)-BPAP – was measured on electrical-stimulation-induced release of labeled transmitters from the isolated brain stem of rats following labeling with [^{3}H]-norepinephrine or [^{3}H]-dopamine or [^{3}H]-serotonin by preincubation in transmitter stores. The study confirmed the selectivity of the enhancer effect of (−)-BPAP (Miklya and Knoll 2003).

Figure 3.10 shows the chemical structure and pharmacological spectrum of tryptamine and two of the tryptamine-derived synthetic mesencephalic enhancer substances.

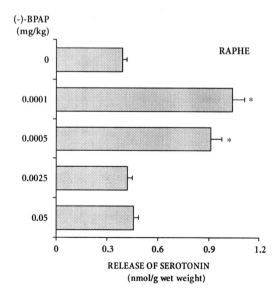

Fig. 3.9. Significant enhancement of serotonin release from the raphe of rats isolated 30 min after the subcutaneous administration of a single dose of (−)-BPAP. The amount of serotonin released from the tissue within 20 min following the administration of different doses of (−)-BPAP was measured according to Knoll and Miklya (1995). *Horizontal lines to the right of the graph bars* show SEM. Paired Student's *t*-test. *$P < 0.01$

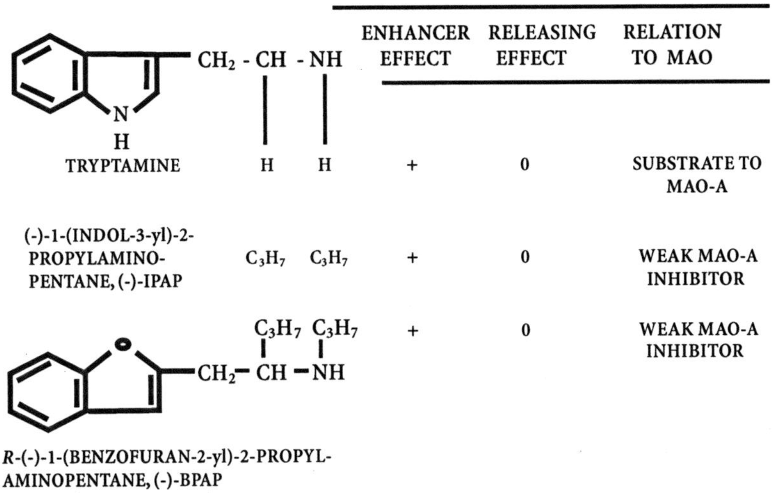

3.2 Pharmacological Analysis of Mesencephalic Enhancer Regulation

			ENHANCER EFFECT	RELEASING EFFECT	RELATION TO MAO
TRYPTAMINE	H	H	+	0	SUBSTRATE TO MAO-A
(-)-1-(INDOL-3-yl)-2-PROPYLAMINO-PENTANE, (-)-IPAP	C_3H_7	C_3H_7	+	0	WEAK MAO-A INHIBITOR
R-(-)-1-(BENZOFURAN-2-yl)-2-PROPYL-AMINOPENTANE, (-)-BPAP	C_3H_7	C_3H_7	+	0	WEAK MAO-A INHIBITOR

Fig. 3.10. The chemical structure and pharmacological spectrum of tryptamine and two of its most representative synthetic derivatives. Taken from Knoll (2001)

3.2 Pharmacological Analysis of Mesencephalic Enhancer Regulation Using (−)-BPAP as a Specific Experimental Tool

3.2.1 Detection of a Specific and a Nonspecific Form of Enhancer Regulation in the Mesencephalic Neurons. Studies Using Isolated Discrete Rat Brain Regions

(−)-BPAP is at present the most selective and potent experimental tool to investigate enhancer regulation in the mesencephalon. The enhancer effect can be detected following the subcutaneous administration of low amounts of (−)-BPAP (see Table 2 in Knoll et al. 1999), as well as following the addition of the substance into the organ bath of freshly isolated discrete mesencephalic brain areas (see Table 3 in Knoll et al. 1999).

Enhancer substances stimulate the enhancer-sensitive neurons in the mesencephalon in a peculiar manner. Figure 3.11 shows the characteristics of the enhancer effect of (−)-BPAP added to isolated locus coerulei of rats. We see two bell-shaped concentration/effect curves. The one in the low nanomolar range, with a peak effect at 10^{-13} M concentration, clearly demonstrates the existence of a highly complex, specific form of enhancer regulation in noradrenergic neurons. The second, with a peak effect at 10^{-6} M concentration, shows the operation of a ten million times less sensitive, obviously nonspecific form of the enhancer regulation in these neurons (see Knoll et al. 2002b, for details).

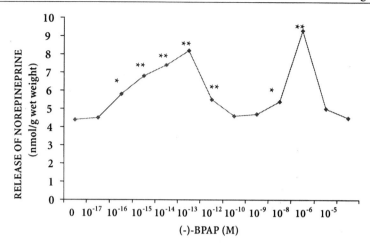

Fig. 3.11. The bi-modal, bell-shaped concentration effect curve characteristic to the enhancer effect of (−)-BPAP on isolated locus coerulei of rats. (−)-BPAP was given to the organ bath of the quickly removed locus coerulei. Eight organs were used for the analysis of each concentration. The amount of norepinephrine released within 20 min from the tissue in the presence of different concentrations of (−)-BPAP was measured according to Knoll and Miklya (1995). Paired Student's t-test. $*P < 0.01$, $**P < 0.001$

We experienced, in a number of studies on rats (Knoll et al. 1955a,b,c, 1956, 1994; Knoll 1956, 1957, 1988) the validity of the common concept that there is a great individual variation in sexual activity and learning performance in any random population of mammals of the same strain. As it will be further discussed later (see Sect. 4.1), the discovery of the bell-shaped concentration/effect curve of the enhancer substance in the low nanomolar concentration range offers the first reasonable explanation for the great individual variation in behavioral performances.

3.2.2
Analysis of the Two Forms of Enhancer Regulation on Isolated Brain Cells in Culture

(−)-BPAP also proved to be a proper experimental tool for detecting the presence and analyzing the nature of enhancer regulation on single brain cells in culture.

Considering the role of the mesencephalic neurons in goal-seeking behavior and collating this experience with the finding that the performance of the catecholaminergic and serotonergic neurons were significantly enhanced in rats *in vivo* with 0.0001 mg/kg (−)-BPAP (see Table 2 in Knoll et al. 1999) and *in vitro* at 10^{-13} M concentration (see Table 3 in Knoll et al. 1999), it

3.2 Pharmacological Analysis of Mesencephalic Enhancer Regulation

was reasonable to assume that this highly sophisticated form of enhancer regulation is the physiological mechanism in the mesencephalon responsible for a drive. We may also assume that from a physiological point of view the enhancement of nerve cell performance elicited by (−)-BPAP in the high concentration range is a non-specific effect, obviously unrelated to behavioral performances.

This view was substantiated by studies with (−)-BPAP on single brain cells in culture: (a) two studies on glial cells (Ohta et al. 2002; Shimazu et al. 2003), (b) a study on mesencephalic neurons (Knoll et al. 1999), and (c) two studies on cortical neurons (Hamabe et al. 2000, and see Figs. 3.12 and 3.13 in the present study).

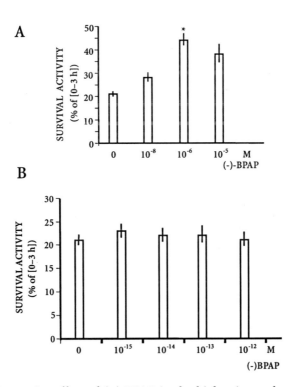

Fig. 3.12. A Protective effect of (−)-BPAP in the high micromolar concentration range, with a peak effect at 10^{-6} M concentration, against serum-free condition induced cell death in low-cell-density culture of the cerebral cortex from E17 rats. B Lack of a protective effect of (−)-BPAP under the same conditions in the low nanomolar concentration range. Experiments were carried out in triplicate. Data are the mean ± SEM from six independent experiments. The data were analyzed using Student's t-test after multiple comparisons of ANOVA. P value was < 0.05 compared with the results in the vehicle-treated culture. See Hamabe et al. (2000) for methodology

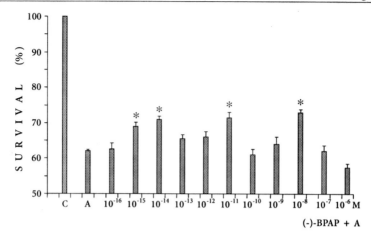

Fig. 3.13. Protective effect of (−)-BPAP against β-amyloid$_{25-35}$-induced cell death on isolated cortical neurons from 8-day-old chicken embryos (Lohman brown hybrid). Duration of a single experiment: 10 days. (−)-BPAP has been added to the culture at the first day *in vitro*. Lesioning with β-amyloid$_{25-35}$ pre-aggregated for at least 72 h. Concentration and stock solution 1 mM, lesioning with 10 µl of the stock solution. *C* control, *A* β-amyloid$_{25-35}$. *Graph bars* given in percent of the unlesioned control (100%), represent the mean viability ± SEM from two independent experiments performed on 2 days with two 96-well plates and six to eight identical wells/concentration and substance. Statistical analysis: two-tailed Student's *t*-test for two means. *$P < 0.05$

3.2.2.1
Studies on Cultured Neuroglial Cells

Neuroglia, the supporting structure of nervous tissue, consists of a fine web made up of modified ectodermic elements in which glial cells are enclosed. Neuroglial cells (astroglia, oligodendroglia, microglia) play an important physiological role in the brain and modulate the function of neurons in a complex manner. They do not participate, however, in the realization of drive-dependent goal-seeking behavior. Thus it was of crucial importance to test the effect of (−)-BPAP on the performance of glial cells. Two studies were performed with (−)-BPAP on cultured mouse astrocytes (Ohta et al. 2002; Shimazu et al. 2003).

As a quantitatively measurable specific function of glial cells, the rate of synthesis of three neurotrophic factors (nerve growth factor [NGF]; brain-derived neurotrophic factor [BDNF]; and glial cell line-derived neurotrophic factor [GDNF]) was measured. In the Ohta et al. (2002) study, the enhancer effect of (−)-BPAP was measured only in the high concentration range. The authors found the amount of NGF, BDNF, and GDNF secreted from astrocytes into the culture medium increased by up to 120, 2, and 7 times more, respectively, than those of the control treatment with 0.35 mM (−)-BPAP for 24 h. The (−)-BPAP-induced increased production of NGF and GDNF was inhibited by concomitant administration of actinomycin D, given for transcription blockade. (−)-BPAP treatment increased the mRNA expression of NGF, BDNF, and GDNF. The results of this study proved that the nonspecific form of the enhancer regulation operates in glial cells.

In the second study the effect of (−)-BPAP was tested in a range of 10^{-15} to 5×10^{-4} M concentration. This study corroborated the finding of the first one. The synthesis of NGF was significantly enhanced in the high micromolar concentration range with a peak effect at 10^{-4} M concentration, whereas 5×10^{-4} M was ineffective. (−)-BPAP acted similarly on the synthesis of BDNF (with a peak effect of 10^{-4} M concentration) and on the synthesis of GDNF (with a peak effect of 10^{-4} M concentration) (Shimazu et al. 2003). But the crucially important step forward was the proof that, as expected, (−)-BPAP was ineffective in the low nanomolar concentration range. Thus the specific form of enhancer regulation was not detectable in glial cells.

This finding supports the view that the specific form of enhancer regulation stimulated by (−)-BPAP in the low nanomolar concentration range is the behaviorally important form, whereas the enhancer effect of (−)-BPAP in the micromolar concentration range is insignificant in behavioral terms. Nevertheless, the (−)-BPAP-induced enhancement of the synthesis of neurotrophic factors is a remarkable pharmacological effect whose therapeutic value deserves further analysis in the future.

3.2.2.2
Studies on Cultured Mesencephalic Neurons

The first analysis of the enhancer regulation on cultured neurons using racemic BPAP as a specific experimental tool was performed on rat hippocampal cells (Knoll et al. 1999).

To elicit cell death the cultured rat hippocampal neurons were treated with β-amyloid$_{25-35}$. BPAP (the racemic substance was used in this early study) exerted its enhancer effect in the characteristic bipolar manner, with bell-shaped concentration/effect curves. The peak effect was reached at 10^{-14} M concentration in the low nanomolar concentration range, and at 10^{-8} M concentration in the higher micromolar concentration range (see Fig. 5 in Knoll et al. 1999). Because of the neurotoxic effect of β-amyloid$_{25-35}$, no more than 20% of the cells, obviously the high performing cells, survived this attack. As the synthetic mesencephalic enhancer substance significantly enhanced the performance of the neurons in culture, in the presence of the optimum concentration (10^{-14} M) of BPAP about 70% of the cells survived.

(−)-BPAP enhanced the activity of the catecholaminergic and serotonergic neurons in isolated discrete mesencephalic regions in the exactly same bipolar manner and in the same concentration range (see Table 3 in Knoll et al. 1999).

The studies with (−)-BPAP performed on noradrenergic, dopaminergic, serotonergic and hippocampal neurons proved unequivocally the operation of a highly specific, complex form of enhancer regulation in the subcortical neurons. This is very much in keeping with the ascription of a commanding role to midbrain neurons in goal-seeking behavior.

3.2.2.3
Studies on Cultured Cortical Neurons

The first study of the enhancer effect on cultured cortical neurons was performed with (−)-BPAP on a primary culture of rat cerebral cortex. In this experiment the rapid cell death of the cortical neurons was measured in serum-free culture. It was shown that in a low-cell-density culture cortical neurons rapidly die. (−)-BPAP, as first shown by Hamabe et al., significantly protected the cortical neurons against serum-free-condition-induced cell death in the high concentration range (see Fig. 2 in Hamabe et al. 2000). The protective effect of (−)-BPAP, with a peak effect at 10^{-6} M concentration, is shown in Fig. 3.12A. However, in striking contrast to the finding on cultured rat hippocampal neurons (see Knoll et al. 1999, Fig. 5), (−)-BPAP did not exert an enhancer effect on the cultured rat cortical neurons in the nanomolar concentration range. This is shown in Fig. 3.12B.

To investigate the difference in sensitivity towards (−)-BPAP between the subcortical and cortical neurons of rats *in vivo*, we performed two series of experiments in the shuttle box.

Tetrabenazine treatment (1 mg/kg s.c.) depletes at least 90% of norepinephrine and dopamine from their stores in the nerve terminals of the catecholaminergic neurons and, due to the weak performance of the catecholaminergic brain engine, the activation of the cortical neurons remains below the level required for the acquisition of a conditioned avoidance reflex (CAR). The learning deficit caused by tetrabenazine treatment can be antagonized by the administration of a synthetic mesencephalic enhancer substance.

In the shuttle box the acquisition of a two-way CAR was analyzed during 5 consecutive days. The rat was put in a box divided inside into two parts by a barrier with a small gate in the middle, and the animal was trained to cross the barrier under the influence of a conditioned stimulus (CS, light flash). If it failed to respond within 5 s, it was punished with an unconditioned stimulus (US), a footshock (1 mA). If the rat failed to respond within 5 s to the US, it was classified as an escape failure (EF). One trial consisted of a 15 s intertrial interval (IR), followed by 15 s CS. The last 5 s of CS overlapped the 5 s of US. At each learning session, the number of CARs, EFs and IRs were automatically counted and evaluated by multi-way ANOVA.

3.2 Pharmacological Analysis of Mesencephalic Enhancer Regulation

To test a compound's ability to enhance the acquisition of CARs in the shuttle box, it is necessary to select proper training conditions. In the case in which the rat was trained with 100 trials per day, the acquisition of CARs reached an 80% level and the EFs approached or reached the zero level. To demonstrate the highly significant enhancer effect of (−)-BPAP on the mesencephalic catecholaminergic neurons *in vivo*, we trained the rat with 100 trials per day, blocked the acquisition of CARs by pretreating the rats with tetrabenazine, and restored the learning ability with the simultaneous administration of (−)-BPAP. Table 3.1 shows that (−)-BPAP antagonized the effect of tetrabenazine in the rats.

Learning is a cortical function and in the series of experiments aiming to test the effect of (−)-BPAP on cortical neurons we trained the rats with 20 trials per day in order to have a chance to detect the drug-induced improvement in learning ability.

Table 3.1 shows that the percentage of CARs in rats trained with 100 trials per day was 77.13 ± 8.47 on the 5th day of the training (Series no. 1). In contrast, it was only 8.50 ± 2.47 in rats trained with 20 trials per day (Table 3.2, Series no. 1).

Table 3.1. Because of its enhancer effect on catecholaminergic neurons, (−)-BPAP antagonized tetrabenazine-induced learning deficit in rats trained in the shuttle box

Series no.	Compound (mg/kg)	Tetra-benazine (mg/kg)	Percentage of CARs	Percentage of EFs	Number of IRs
1	Saline	–	77.13 ± 8.47	6.00 ± 5.72	34.25 ± 11.21
	(−)-BPAP				
2	–	1	5.00 ± 3.30	61.50 ± 13.80	5.83 ± 2.18
3	0.05	1	46.88 ± 14.15*	17.88 ± 9.30***	9.25 ± 2.81
4	0.10	1	46.38 ± 8.75***	7.38 ± 4.34****	6.75 ± 1.03
5	0.25	1	59.00 ± 12.62***	5.25 ± 2.13****	16.75 ± 5.74
6	0.50	1	70.38 ± 10.73****	1.38 ± 1.02****	8.50 ± 2.83
7	1.00	1	87.75 ± 1.95****	0.13 ± 0.13****	27.38 ± 4.49**
8	2.50	1	79.75 ± 7.03****	1.38 ± 1.12****	24.50 ± 9.19*
9	5.00	1	92.00 ± 2.47****	0.00****	57.88 ± 19.37*
10	10.00	1	92.00 ± 2.46****	0.00****	68.33 ± 26.46*

Tetrabenazine or the combination of tetrabenazine + (−)-BPAP was administered subcutaneously, 60 min before daily measurement.
Rats (in each group 4 males and 4 females) were trained at 100 trials daily for 5 days in the shuttle box. The performance on the fifth day of training is shown in the table.
CAR conditioned avoidance response; *EF* escape failure; *IR* intersignal reaction
Significance of combination (tetrabenazine + (−)-BPAP) vs tetrabenazine (ANOVA):
*$P < 0.05$, **$P < 0.02$, ***$P < 0.01$, ****$P < 0.001$.

Thus, in case (−)-BPAP had possessed a specific enhancer effect on cortical neurons, we could detect it easily in form of a significant, dose-dependent increase in the percentage of CARs and in the reduction of the percentage of EFs.

Because of the bell-shaped concentration effect curve characteristic to the enhancer effect of (−)-BPAP (see Fig. 3.11), we used 10 doses of the compound, ranging from 0.000001 to 10 mg/kg, to clarify the effect of (−)-BPAP on the cortical neurons. Table 3.2 demonstrates that none of the applied doses of (−)-BPAP was capable of changing the learning performance of rats in the shuttle box. In accord with the findings on cultured rat cortical neurons, the *in vivo* experiments confirmed that (−)-BPAP, the presently known most potent synthetic mesencephalic enhancer substance, is devoid of a specific enhancer effect on the cortical neurons.

The second study of the enhancer effect on cultured telencephalic neurons was performed on cortical cells from 8-day-old chicken embryos (Lohman brown hybrid). This is the only study to date on nonmammalian brain cells. (−)-BPAP detected the operation of both the specific and nonspecific form of enhancer regulation in the cortical neurons of this avian species. The performance of the cortical neurons was enhanced in the low nanomolar concentration range of (−)-BPAP, with a peak effect at 10^{-14} M concentration (Fig. 3.13).

Table 3.2. Because of its ineffectiveness on cortical neurons, (−)-BPAP did not enhance the learning performance of rats trained in the shuttle box

Series no.	(−)-BPAP (mg/kg)	Percentage of CARs	Percentage of EFs	Number of IRs
1	Saline	8.50 ± 2.47	0.75 ± 0.62	1.88 ± 1.01
2	0.000001	6.13 ± 1.99	1.25 ± 0.65	4.00 ± 2.65
3	0.00001	4.38 ± 2.08	3.25 ± 1.56	3.63 ± 1.46
4	0.00005	5.88 ± 2.44	2.38 ± 1.96	2.50 ± 1.02
5	0.0001	12.75 ± 2.18	0.63 ± 0.63	1.25 ± 1.68
6	0.0005	9.63 ± 2.07	0.63 ± 0.42	6.50 ± 3.08
7	0.025	8.50 ± 2.48	2.25 ± 1.16	2.88 ± 1.39
8	0.05	8.63 ± 2.13	0.00	4.38 ± 2.34
9	0.1	6.75 ± 2.96	2.13 ± 2.13	2.13 ± 0.61
10	1.0	8.50 ± 2.63	0.00	4.25 ± 1.82
11	10.0	0.63 ± 0.42	1.13 ± 0.74	2.88 ± 1.26

(−)-BPAP was administered subcutaneously, 60 min before daily measurement Rats (in each group 4 males and 4 females) were trained at 20 trials daily for 5 days in the shuttle box. The performance on the fifth day of training is shown in the table.
CAR conditioned avoidance response; *EF* escape failure; *IR* intersignal reaction
Significance of (−)-BPAP vs saline was calculated according to ANOVA; in all cases $P > 0.05$.

For a trial aiming to explain the striking sensitivity difference of the cultured cortical cells of rats and chickens towards (−)-BPAP see Sect. 3.4.1.

3.3
Considering Enhancer Receptors

According to our present knowledge substances that change the activity of a cell in very low concentrations exert their effect via a highly specific receptor. The finding that (−)-BPAP enhances the activity of the noradrenergic, dopaminergic, serotonergic and hippocampal neurons in the brain at 10^{-13}–10^{-16} M concentration (Knoll et al. 1999) speaks in favor of the existence of highly specific enhancer receptors in these neurons.

In order to get direct evidence for (−)-BPAP-sensitive receptors, we performed experiments with [^{3}H]-(−)-BPAP. Unfortunately, we were unable to find unequivocal evidence for the predicted binding sites for [^{3}H]-(−)-BPAP. Furthermore, the explanation of the peculiar bell-shaped concentration/effect curve characteristic of the enhancer effect of (−)-BPAP (see Fig. 3.11) remains to be understood.

It is obvious that the enhancer effect of (−)-BPAP exerted, for example, on the isolated locus coeruleus of rats (see Fig. 3.11) with a peak at 10^{-13} M concentration is a highly specific effect from a physiological point of view, while the second peak at 10^{-6} M concentration represents a nonspecific effect of minor importance. The proposition that (−)-BPAP binds to a specific receptor in the technically unmeasurable concentration range and higher concentrations induce a conformational change which makes the binding of the ligand impossible would explain both the bell-shaped concentration/effect curve and the ineffectiveness of our trials to furnish unequivocal evidence for the binding of [^{3}H]-(−)-BPAP to its receptor. Nevertheless, the observed phenomena need clarification.

Because we were unable to find, using [^{3}H]-(−)-BPAP, direct evidence for specific enhancer receptors, we tried to approach the problem from another angle. In the rat brain, using a classic pharmacological method, we found convincing indirect evidence for (−)-BPAP-sensitive enhancer receptors in the mesencephalon (Knoll et al. 2002a).

1-(2-Benzofuryl)-2-(3,3,3-trifluoropropyl)-aminopentane HCl (3-F-BPAP), a close structural analogue of BPAP with weak enhancer activity, was synthesized with the expectation that the simultaneous administration of this analogue with (−)-BPAP would significantly antagonize the enhancer effect of the latter, proving that they act on the same receptor. The low specific activity of 3-F-BPAP was demonstrated in the rat in the shuttle box.

The subcutaneous administration of 1 mg/kg tetrabenazine depletes the catecholamine stores in the brain within 1 h. As a consequence of this change, tetrabenazine treatment inhibits the acquisition of a two-way avoidance reflex in the shuttle box. This effect can be significantly antagonized by enhancer substances. The effect of (−)-BPAP was measured in eight different doses from 0.05 to 10 mg/kg. Even the lowest dose significantly antagonized tetrabenazine-induced inhibition of learning (see Table 3.1). In contrast, 3-F-BPAP was ineffective in five different doses, ranging from 0.25 to 5.0 mg/kg (Table 3 in Knoll et al. 2002a).

The concurrent administration of 1 mg/kg 3-F-BPAP with 0.1 mg/kg (−)-BPAP significantly inhibited the enhancer effect of (−)-BPAP, but 1 mg/kg 3-F-BPAP did not influence the enhancer effect of 1 mg/kg (−)-BPAP (Fig. 2 in Knoll et al. 2002a). This is a clear indication that the compounds bind to the same receptor, to which (−)-BPAP has a much higher affinity than 3-F-BPAP.

(−)-Deprenyl, at present the only enhancer drug in general use, though being substantially less potent in the shuttle box than (−)-BPAP, significantly antagonized the learning deficit caused by tetrabenazine. We studied the effect of 1 and 5 mg/kg (−)-deprenyl in different combinations with 1 and 5 mg/kg 3-F-BPAP and found that 3-F-BPAP left the enhancer effect of (−)-deprenyl unchanged (Fig. 3 in Knoll 2002a). Furthermore, 3-F-BPAP did not influence the enhancer effect of (−)-PPAP, the (−)-deprenyl analogue free of MAO-B inhibitory potency (Fig. 4 in Knoll 2002a).

The data prove that the molecular mechanism through which the PEA-derived substances, (−)-deprenyl and (−)-PPAP, exert their enhancer effect *in vivo* is not identical with the mechanism through the stimulation of which the tryptamine-derived substance, (−)-BPAP, acts. This is in accord with the finding that, in contrast to (−)-BPAP, (−)-deprenyl did not exert an enhancer effect on the serotonergic neurons (see Fig. 3.5, and for more details Knoll et al. 1999). That (−)-BPAP enhances the activity of the catecholaminergic and serotonergic neurons in the rat brain via the stimulation of a highly specific enhancer receptor is strongly supported by the finding that the compound did not show a significant binding capacity to any of the receptors known to play a role in the function of the catecholaminergic and serotonergic neurons (see Table 3 in Knoll et al. 1999).

The characteristic enhancer effect of (−)-BPAP, as shown for example in Fig. 3.11, in the low nanomolar range and at a higher micromolar range (Knoll et al. 1999, Yoneda et al. 2001) indicate the existence of two types of (−)-BPAP-sensitive enhancer receptors in the brain stem neurons represented by high- and low-affinity binding sites. The recent identification of a family of G-protein-coupled trace-amine receptors in the mammalian brain specifically stimulated by the endogenous enhancer substances, PEA and tryptamine (Borowsky et al. 2001), strongly suggests that the authors located a family of enhancer receptors. This assumption is suggested by the finding that amphetamine and metabolites of the catecholamines neurotransmitters were also found to be antagonists of a rat trace-amine receptor (Bunzow et al. 2001).

3.4 Cortical Enhancer Regulation

The obvious difference already established between the binding of (−)-deprenyl and (−)-BPAP (Knoll et al. 1999) argues for the existence of various types of enhancer receptors. Remarkably, Borowsky et al. (2001) found that more than one member of the newly identified family of mammalian G-protein-coupled receptors was activated by PEA and tryptamine.

Studies with (−)-BPAP, the most potent and selective synthetic mesencephalic enhancer substance, which is presently also the best experimental tool for the analysis of the binding of enhancer substances to receptors, are at the very beginning. Nevertheless, two studies have already been published showing that (−)-BPAP has remarkable binding capacity to some receptors. Hamabe et al. (2000) demonstrated that high concentrations of (−)-BPAP displaced the binding of $[^3H]$-(+)-pentazocine to sigma receptors in the synaptic membranes from rat cerebral cortex. Thereafter, Rashid et al. (2001) has found that (−)-BPAP binds to metabotropic sigma receptors in peripheral nociceptor endings. The sigma agonist-induced nociception was found to be due to the release of substance P from nociceptor endings through activation of $G\alpha_{i1}$ and phospholipase C (Ueda et al. 2000). A number of studies indicated that sigma agonists stimulate heterometric G-proteins (Connick et al. 1992; Tokuyama et al. 1999; Maruo et al. 2000). The nociceptive flexor responses in mice induced by both (+)-pentazocine and (−)-BPAP were blocked by sigma receptor antagonist BD 1063. In radio-ligand binding assay, $[^3H]$-(+)-pentazocine showed a saturable specific binding in membrane preparation from mouse liver, and this specific $[^3H]$-(+)-pentazocine binding was inhibited by (−)-BPAP as well as by (+)-pentazocine and BD 1063 (Rashid et al. 2001). According to our present knowledge it is hard to find any reasonable relation between the binding of (−)-BPAP to the sigma receptors and its enhancer effect.

Not only the real nature of the specific mesencephalic enhancer receptors but also the endogenous ligands to these receptors remain unresolved. The high potency of (−)-BPAP in comparison to the already identified natural enhancer substances, PEA and tryptamine, is remarkable. This difference gives justification for the search for much more potent natural enhancer substances than PEA and tryptamine.

3.4 Cortical Enhancer Regulation: Assumptions About Its Physiological Significance

3.4.1 Essential Forms of the Modification of Behavior Through Exercise, Training, or Practice

The remarkable difference in sensitivity for (−)-BPAP between the isolated cortical cells of rat and chicken (see Sect. 3.2.2.3) is worthy of particular attention. (−)-BPAP exerted its enhancer effect with a peak at 10^{-14} M concentration on isolated cortical neurons from 8-day-old chicken embryos, but was ineffective in the nanomolar concentration range on a primary culture of rat cerebral cortex. Although the experimental tool used for eliciting cell-death was β-amyloid$_{25-35}$ in the chicken experiment (Fig. 3.13) and serum-free condition in the rat study (Fig. 3.12), it seems unreasonable to make the applied methods responsible for the observed difference. It seems much more plausible that we are dealing with a basic functional difference between the cortical neurons of the two species: *rats possess the ability to acquire drives, chickens are devoid of it.*

The faculty for acquiring a drive is uncommon in the animal kingdom. It was shown by Berta Knoll in the late 1950s that the mouse, a rodent closely related to the rat, was unable to acquire the glass-cylinder-seeking drive (B. Knoll, Thesis, 1968). She has found that, in striking contrast to the rat, the mouse was even unable to fix the inextinguishable form of the CAR, the functional stage that preceded the acquisition of the glass-cylinder-seeking drive in the rat (B. Knoll 1961).

In the initial training phase leading to the manifestation of the glass-cylinder-seeking drive, the rat was forced – for a couple of weeks, three times daily, on 10–50 occasions – to jump, when pushed through the side opening of a glass cylinder standing on a metal plate heated to 60 °C, on to the upper rim of the glass cylinder. The rat's behavior was modified after a short training period. The animal soon escaped from the unheated plate within 10 s, even 100 times in succession. The acquired inextinguishable CAR remained stable without reinforcement.

Trained under the same experimental conditions, the mouse seemed to behave similarly to the rat. To the end of the daily experiment the mouse escaped from the unheated plate within 10 s, even 100 times in succession. However, in striking contrast to the rat, the mouse was unable to fix the acutely detectable modification of behavior in the cortex. The next day there was no sign of the acquisition of a CAR. Even weeks of daily training did not modify this behavior of the mice. Thus, the experiments furnished unequivocal evidence for a qualitative difference in the natural endowments of the cortical neurons of mice and rats. The rat brain possesses the ability to fix chains of ICRs, while the mouse brain did not reach this stage of development.

From a physiological point of view the brains of members of the same strain are undeniably equal in their natural endowments. The device is the same. In the human, however, the extreme differences in life conditions that primarily determine the realm of the acquired drives, combined with substantial individual differences in learning ability, make unpredictable as to which trifling proportion of the immense inborn functional pool will be utilized. An individual necessarily strives to build those forms of acquired drives that demand the shortest training time with the lowest investment of energy. It is the plastic

3.4 Cortical Enhancer Regulation

description of this phenomenon that if their life conditions allow it, individuals aspire to select their activities according to their abilities.

To translate this description into the language of neurochemistry, we may say that it is the natural endowment of individuals to give preference to activities according to the efficiency of the enhancer regulation in the population of cortical neurons responsible for the selected performance. The best performing, the *talented* individuals will be the ones who mobilize when needed the specific enhancer substance at the optimum concentration (see Sect. 3.2.1).

We compared, in our longitudinal studies on rats, innate (hunger and sexual) drives with an acquired (glass-cylinder-seeking) drive. We arrived at the conclusion that the modification of behavior through practice, training, or experience means that groups of neurons acquire the ability to change the functional state of other groups which thereafter enables their cooperation. The sequence of learning-induced, exactly measurable modifications of behavior in the rat allows us to define four functionally different states of cortical neurons, the chemistry of which remains to be clarified. Accordingly we distinguish four functionally different groups of cortical neurons:

Group 1. Cortical neurons in their inborn naive state. Neurons are born to perceive special senses (light, color, sound, smell, taste, pain, touch) and when stimulated, an evoked potential is detectable in the especially sensitive group of cortical neurons within 0.015 s. The mammalian brain is supplied with such a high number of cortical neurons that, considering the short lifetime of the organism, a high percentage of the cortical neurons probably preserve their inborn functional state.

Group 2. Cortical neurons whose functional state has been modified through learning to serve an ECR. Pavlov's discovery that the main condition for the acquisition of an ECR is the stimulation of a special group of cortical neurons with their specific stimulus *simultaneously* with the precipitation of an unconditioned reflex can still hardly be overestimated. This type of behavioral modification has certainly been the most thoroughly studied phenomenon in the history of brain research. In our experiments the physiological significance of this functional state was illustrated by the quick adaptation of the glass-cylinder-seeking rats, which generate and extinguish long chains of ECRs (tool reflexes) according to need, enabling them to easily reach the goal in spite of unforeseeable changes in the environment (Knoll 1969). Mainly *followers of Pavlov* tried for a while to maintain the false doctrine that the totality of higher nervous activity can be explained through this mechanism.

Group 3. Cortical neurons whose functional state was modified through learning to serve an ICR. We experienced the fixation of chains of ICRs in the course of the training procedure aiming to develop the glass-cylinder-seeking drive in rats. It was shown in Daniel Bovet's laboratory by Kelemen et al. (1961) that an ICR can sharply be differentiated by EEG from an ECR. It seems obvious that in every moment of life the already firmly fixed, readily ecphorizable stock of chains of ICRs represents the consciously perceivable, stable knowledge of the individual.

Group 4. Cortical neurons whose functional state was modified through learning to serve an acquired drive. A population of cortical neurons (the cortical representation of the drive) acquires, through proper training, the ability to produce, when needed, the specific enhancer substance in an optimum concentration, reach the highest possible level of excitability and stay in this state continuously ('active focus') until the goal is reached. Drives determine the life of the mammalian organisms as their operation is the condition *sine qua non* for the building, fixing, and ecphorizing chains of ECRs and ICRs whenever required.

For a domesticable mammal ready to acquire drives, life means the practice-, training-, or experience-induced continuous transition of the functional state of cortical neurons in the above-cited sequence from Group 1 to Group 4. As a consequence of these changes, we observe the proper modification of behavior. The human brain is, of course, the best model for studying this chain of events.

*

A healthy human brain possesses, practically speaking, an immense capacity to fix chains of ICRs and acquired drives. This is obvious when we call to mind, as an example, the fact that the human brain is born with billions of neurons belonging to Group 1 that just serve the sense of hearing. If we follow this train of thought and restrict ourselves only to one aspect of this basic brain function, the ability to fix chains of ICRs belonging to the world of music, it is easy to realize that this primarily hearing-dependent capacity of the human brain is by itself inexhaustible. And this is all the more so, since the chains of ICRs can also be fixed by reading the proper notes.

It depends of course on living conditions which small proportion of disposable neurons will definitely ascend in the hierarchy and thus assume the role of Group 3 neurons during the lifetime of the individual. We know that whatever human activity is measured we observe extreme individual differences in performance due to substantial individual differences in the cortical enhancer regulation (see Sect. 4.1). But, in the course of the short human lifetime, even a talent on the order of magnitude of a Johann Sebastian Bach can utilize only a humble proportion of the neurons available for fixing chains of ICRs in the field of music. This is true of all kinds of human activities, since the human brain possesses a network of over 100 billion interrelated nerve cells and a 10^{10} bit capacity.

*

A human cortex is, at birth, comparable with a book consisting of billions of empty pages. Life is obviously too short to scribble over this entire book. Every mammalian organism can be defined at any given moment by the number of

3.4 Cortical Enhancer Regulation

neurons that have already changed their functional state by this date. In the case of human beings the self is determined by the already fixed chains of ICRs and acquired drives, as their operation, in contrast to the function of chains of ECRs, is inseparable from conscious perception.

Domestication of animals proves that even in ancient times humans had recognized the ability of some animal species to acquire drives for unnatural goals and made a good use of it. It is reasonable to assume that the essential cortical mechanism responsible for the transition of a naive neuron in a sequence of events until the acquired drive is fixed is the same when a rat acquires the glass-cylinder-seeking drive or a human fixes any form of an acquired drive. Nevertheless, there can be no doubt that human performance is qualitatively different from the performance of a domesticable animal. We may interpret this difference as a typical example of the transition of quantity into quality. There is an enormous quantitative difference between the most clever animals and humans in the ability of their naive cortical neurons to change their functional state and ascend in the hierarchy until they ultimately become part of Group 3 or 4. In striking contrast to the cortex of the domesticable animals, the human cortex is capable of making this alteration with ease and high speed. Compared to humans, the ability of the brain to fix chains of ICRs and to acquire a new drive is a rudimentary function even in the anthropoid apes.

In our studies aiming to build the glass-cylinder-seeking drive into the brain of rats, we trained hundreds of animals and followed their performance until they died (Knoll et al. 1955a,b,c, 1956; Knoll 1956, 1957). In one series of experiments on a random group of 100 two-month-old rats (50 males, 50 females) we developed – within a 3-week training period in each animal – the inextinguishable conditioned jumping reflex. Yet out of the hundred rats, only 20% of the population (11 females, 9 males) showed clear-cut signs of a tendency to acquire the glass-cylinder-seeking drive and only two of them (one male and one female) ultimately maintained this drive through their entire life. If we compare the restricted ability of rats to build acquired drives to the almost unlimited ability of the human brain to acquire new drives, the qualitative difference in performance is understandable without any need to deny that the mechanism of the cortical enhancer regulation is essentially the same in the two species. This is simply a new example of the general rule in nature that an immense variety of colorful phenomena, in this case the fantastic variation in the outward form of behavioral performances, rests on the operation of a common simple mechanism. Somewhat like "Gravitation keeps the whole universe going."

Although any form of an acquired drive is rooted in one of the innate drives, as soon as the new drive develops and operates in an inextinguishable manner, the roots become unrecognizable. Watching a glass-cylinder-seeking rat in operation, one cannot recognize that escape from a hot plate was the foundation of this acquired drive.

It seems reasonable to assume that, from a functional point of view, the appearance of species with the ability to acquire drives for unnatural goals was the last radical turning-point in the development of brain organization. In the animal kingdom the new mechanism reached its functionally most sophisticated level in the group of anthropoid apes, but it reached perfection in *Homo sapiens* only. The new mechanism culminated in the development of speech – the classic, human-specific instrument that made interpersonal communication possible by capturing reality in the form of symbols – and thus opened the way for the operation of an unrestricted variety of acquired drives.

The learning of each letter is in itself the fixation of a chain of ICRs. The learning of each word represents the fixation of a much more complicated chain of ICRs in which the sequence of letters is determinant. The words will then be used as tools to build sentences. A sentence is part of an acquired drive induced goal-oriented behavior. The words are used as rapidly changeable tools for reaching a practically infinite number of goals. Speech, in conjunction with all other forms of language-based interpersonal contacts, brought forth the most sophisticated technique in service of goal-oriented behavior and produced the highest level of human achievements, science and art. This method allowed, with greater or lesser efficiency, the preservation of the achievements of ancestors, leading to the uniqueness of human society. In this society each generation stands on the shoulders of past generations, evaluating history and envisaging the future.

*

To understand the past and envisage the future of the human society we should never forget that the last step forward in the development of life on earth, the evolution of brains with the ability to acquire drives for goals unnecessary for the survival of the individual or the species, is based on a cortical mechanism that humans share with a couple of animal species. An "active focus" is created in the brain through learning: a population of neurons ascend into Group 4.

The operation of the acquired-drive-directed behavior means, objectively, the proper chemical changes in the cortical representation of the drive, subjectively, the imagination (mental representation) of the goal to be reached. As a matter of fact, the operation of those neurons in the brain that have learned to cooperate with each other and represent an integral whole also represents the proper cognitive/volitional and affective consciousness which we simply describe as the imagination of the goal to be reached. In practice, when the specific enhancer substance is produced in its optimal concentration and keeps the neurons belonging to the "active focus" at their highest level of excitability, the urge starts operating and the individual is ready to conquer any obstacle to reach the subjectively imagined goal.

From the point of view of the basic physiological mechanism of acquired-drive-directed behaviors there is no difference between a glass-cylinder-

3.4 Cortical Enhancer Regulation

seeking rat and a scientist or artist who seeks to reach a special, highly sophisticated goal. *Only the goals to be reached are qualitatively different.* Using the same mechanisms, the rat is looking for a glass cylinder, the creative scientist for something previously unknown, and the creative artist for something previously nonexistent.

The creative human mind is the best example for understanding the essential characteristics of acquired-drive-directed behavior. For sake of illustration we may take as an example an immortal achievement of the human brain, the unusually well-documented birth of *Composition VII*, the most monumental oil painting by Kandinsky, a master for whom the creation of a work of art was the creation of a world.

Kandinsky executed the painting between November 25 and 28, 1913. His pupil and life-partner, Gabriele Münter, took four photographs in the span of 3 1/2 days, documenting the progress of the work. On the other hand, 33 works (drawings, watercolors, oil paintings) related to this composition are known. Kandinsky, who wished to demonstrate that color is as expressive and powerful as sound in making art without narrating anything realistic, irrefutably proved his thesis with *Composition VII* (first exhibited in 1914 in Cologne). This is truly a breathtaking symphonic construction in painting.

Collating the analysis of the 33 related works (Dabrowsky 1995) with Gabriele Münter's photos showing the progress of the work until the oil painting was finished, we see, in a highly sophisticated form, the trial and error mechanism that always operates until a goal is reached. The acquired drive, the "active focus", subjectively the imagination of the glass cylinder, drove our rats until the goal was reached; and the acquired drive, the "active focus", subjectively the imagination of *Composition VII*, drove Kandinsky until his goal (now permanently exhibited in the Tretyakov Gallery in Moscow) was reached.

Basic laws are simple, gray. The phenomena brought into existence by them are, however, immensely complex and colorful. With the billions of functional units in the brain capable of cooperating with each other, a simple mechanism, cortical enhancer regulation, brings into existence an immense variety of colorful acquired-drive-directed behaviors. Glass-cylinder seeking is an example of the simplest forms of such behavior, while production of *Composition VII* represents one of the most sophisticated, breathtakingly complex forms of acquired-drive-produced behaviors.

3.4.2
The Concept that Learning Is a Cortical Enhancer Regulation Dependent Function

In vertebrates, learning – the modification of behavior through practice, training, or experience, one of the essential necessities of life – is the main physiological function of the cortex. Modification of behavior rests upon the inborn ability of cortical neurons to get acquainted with each other through training, learn to influence each other's function, and cooperate thereafter according to need. The mechanism of this important process is, however, still unknown. The discovery of enhancer regulation offers the following interpretation of learning.

Each member of a population of naive cortical neurons (Group 1) born to perceive a specific quality of stimuli, originating from outside or inside the body, synthesizes the same enhancer substance. It is also supplied with enhancer receptors for which this enhancer substance is the highly specific ligand. The stimulation of the neurons with their enhancer substance leads to enhanced excitability. On the other hand, each cortical neuron is able to activate under proper conditions (training) an enhancer receptor to any of the existing cortical enhancer substances (learning). Thus, *neuron A is born with its specific enhancer receptor* (ER_A) *and with the ability to synthesize its own enhancer substance* (ES_A). *Neuron B is born with* ER_B *and synthesizes* ES_B, *and so on. Whenever a cortical neuron gets excited, its specific enhancer substance is synthesized in an increased amount, and its sensitivity toward other enhancer substances is significantly increased.*

When neurons A and B are simultaneously stimulated, both are continuously bombarded with a higher amount of the enhancer substance of the other neuron and at the same time also sensitized to activate a receptor to the alien enhancer substance. As a consequence, the concurrent stimulation of neurons A and B time after time (training) ultimately leads to the fixation of a new functional constellation. Neuron A acquires sensitivity toward ES_B, and neuron B acquires sensitivity toward ES_A. Thus, learning means that a neuron acquires the ability to respond to originally alien stimuli. As a consequence of this change we experience the training induced modification of behavior.

Using the shuttle box technique, there is a reasonable possibility of testing the validity of this concept on rats. The shuttle box is a simple and useful setup for following the development of a two-way conditioned avoidance reflex (CAR). The box is divided inside into two parts by a barrier with a small gate. The rat is trained to cross the barrier under the influence of flash light (conditioned stimulus, CS). If the rat fails to do so, the animal is punished with an electric footshock (unconditioned stimulus, US). The rat is trained with 100 trials/day. One trial consists of a 15 s intertrial interval, followed by 15 s flash light that overlaps with a footshock for 5 s. If the rat does not cross the barrier to footshock within 5 s, this is noted as an escape failure (EF). The rat learns to avoid punishment and escapes in response to flash light within 10 s (CAR). The percentage of CARs and EFs as well as crossings during the 15 s intertrial interval (intertrial response, IR) is automatically registered.

According to present views, the rat, driven by fear, tries to prevent punishment and learns by trial and error to escape in due time. The efficiency of learning is thought to be proportional to the number of the successful crossings in response to flash light within 10 s. According to our new concept the

3.4 Cortical Enhancer Regulation

efficiency of learning depends on the repeated simultaneous operation of functionally different populations of cortical neurons. In the light of this approach we need to weigh carefully the series of events in the cortex during the training procedure.

The concept predicts that the development of a stable CAR in the shuttle box signifies the acquisition of a special cooperation between the groups of cortical neurons born to perceive the footshock (US) and the flash light (CS), respectively. Nevertheless, other groups of cortical neurons (stimulated, e.g., by the setup as a whole) are also involved in the special modification of the rat's behavior. In the course of training numerous groups of cortical neurons, A, B, C... n, born to perceive special information only, are synchronously active and influence each other. Furthermore, each group of neurons has the chance to develop sensitivity toward each of the enhancer substances belonging to the simultaneously activated groups of neurons. Thus, during the training procedure a network of cooperating groups of cortical neurons develops, which operates thereafter as an entity. The training-induced cooperation between the groups of neurons can be 1. transient in nature (chain of ECRs), 2. irreversibly fixed (chain of ICRs), or 3. may lead to the development of the most sophisticated form of excitatory state in a group of cortical neurons ("active focus") that will operate thereafter as an acquired drive. However complicated the cooperation developed between different group of neurons during training may be, it is their common feature that they work thereafter as an integral whole, and this entity can be activated via a few decisive groups of neurons.

Humans, capable of communicating via symbols, can easily experience the operation of this mechanism. Each letter, a basic symbol of communication, is by itself a chain of ICRs fixed forever in the brain when reading/writing was learned. Each word is a much more complicated chain of ICRs based on the special sequence of letters. A sentence is an acquired-drive-induced goal oriented function that uses words as tools.

A word consisting of, let's say, eight letters is a chain of eight chains of ICRs that are fixed in a special sequence, and the totality of this complicated system is perceived as a whole. Whenever the word is ecphorized as a whole, this means the explosion-like activation of the groups of cortical neurons in the sequence as they were fixed when we learned the word.

As it will be discussed in detail later (see Sect. 4.2), it is enough to see the first, last and one or two intermediate letters to activate the whole chain of the irreversibly fixed eight chains of ICRs in their special sequence and perceive the word as an entity *with the same speed* as the correctly written one. If these letters are left untouched, all the words of a long sentence can be similarly misspelled without interfering with comprehension of meaning. In case of a higher degree of confusion of the letters, a longer time is needed until the word as an integral whole can be consciously ecphorized. Based *on the experience* of this well-known ability of the human cortex, it has become a popular TV quiz all over the world to present the completely jumbled letters of a longer word or a short sentence, with the prize going to the contestant who most quickly recognizes the word or phrase.

According to our approach learning means the establishment of a cooperation between functionally different groups of cortical neurons (Group 1), born to perceive only one special type of stimulation. In response to proper training, however, they also learn to respond to an alien stimulus. Thus whenever we ecphorize the acquired engram, we simultaneously activate functionally heterogeneous groups of cortical neurons. It depends on the quality of training whether an untrained cortical neuron (Group 1) ascending in the hierarchy assumes the role of those of Group 2 (ECR) Group 3 (ICR), or Group 4 ("active focus"), but only the proper activation of Group 3 and 4 neurons is inseparable from conscious perception.

It is obvious that much more time is needed for the activation of functionally heterogeneous groups of cooperating neurons than for the activation of a functionally homogeneous group of neurons. This has already been unequivocally proven in humans (Libet 1973), although the reason for the observed phenomenon remained unexplained until now.

Libet performed a series of experiments on fully conscious patients during the exposure of a cerebral hemisphere for some neurosurgical procedure. The postcentral gyrus was electrically stimulated with extreme care in order to establish the conscious perception of this stimulation.

By definition, only the proper stimulation of neurons belonging to Group 3 or 4 is inseparable from conscious perception. Here we have to remember even an excellently operating glass-cylinder-seeking rat that was brought to the laboratory every day lingered for a longer period of time before it started to work. We described this phenomenon as "warming up" (see Knoll 1969, for review). It seems reasonable to assume that a longer time is needed until the enhancer regulation in the cortical neurons belonging to Group 4, that operate as the "active focus," "the cortical representation of the drive," is transformed to the state at which production of the enhancer substance reaches the critical concentration and the neurons arrive at the level of excitability, subjectively to the "imagination of the glass cylinder," which is the precondition for the readiness to go through fire and water to reach the goal.

Protocols 1, 2 and 3 (Sect. 4.2) provide examples of the "warming-up" phenomenon. Protocol 1 shows, for example, that on March 4 the rat reached the goal in 12'01" at the first trial, in 4'29" at the 2nd trial, and this time varied between 57" and 2'35" in the following eight trials. The rats whose performance is registered in Protocols 2 and 3 behaved similarly.

Libet found that a single stimulus was ineffective. For the conscious perception of cortical stimulation he needed to apply trains of 0.5-ms pulses at liminal intensity for as long as 0.5-s duration.

3.4 Cortical Enhancer Regulation

Thus Libet has detected experimentally in humans the phenomenon we observed in rats and described as "warming up." He just did not know what happened. He obviously activated, with the aid of electrical stimulation, a chain of ICRs and ecphorized the engram as an integral whole. He needed 0.5 s until the cooperating neurons were brought to the state of excitability inseparable from conscious perception.

Libet found that the same long period was required for the sensory perception of a cutaneous stimulation. He stimulated the skin of the hand with a brief electrical pulse. Though 0.015 s is enough to elicit an evoked potential in the somesthetic cortical area in response to stimulation, *once again 0.5 s elapsed before the skin stimulation was consciously perceived.*

This means that in the Libet experiment a period of time 33 times longer (0.5 s) was needed for the activation of the functionally heterogeneous groups of cooperating cortical neurons that learned to work together in the past, ascended in the hierarchy and assumed the role of Group 3, than for the activation of a functionally homogeneous group of naive cortical neurons (0.015 s).

With all this in mind, our approach was that the modification of behavior of the rats trained in the shuttle box depends on the synchronous activation of different groups of cortical neurons in the brain for a proper period of time. The following method is suitable to test the validity of this approach.

Treatment of rats with 1 mg/kg tetrabenazine, which blocks selectively and reversibly the reuptake of the catecholaminergic transmitters into their intraneuronal stores, depletes norepinephrine and dopamine from the end organs of the catecholaminergic neurons in the brain stem. Since the operation of the catecholaminergic brain engine is the condition *sine qua non* for the trial-and-error mechanism and thus for the success of reaching a goal, the acquisition of a CAR in the shuttle box cannot be detected in tetrabenazine-treated rats because of the blockade of the animal's ability to cross the barrier.

Nevertheless, the activation of the cortical neurons via the US and CS remains unchanged in tetrabenazine-treated rats, as can be shown by measuring the evoked potentials following stimulation in the proper cortical area. Thus, according to our concept, the condition in the shuttle box for learning must be unchanged in tetrabenazine-treated rats. To be able to detect whether rats learn when pretreated with tetrabenazine, we performed the experiments on Charles River Wistar Wistar rats which, according to our earlier experiments, proved to be a strain with exceptionally low learning capacity in the shuttle box (Knoll et al. 1996c). We used females as they are even worse performers than males.

In the first part of the experiment (Series A), groups of female rats ($n = 8$) were trained in the shuttle box from Monday until Friday with 100 trials/day for 5 consecutive weeks. The animals were treated subcutaneously, 1 h prior to measurement, either with 1 ml/kg saline (Group 1) or with 1 mg/kg tetrabenazine (Group 2). Following a 5-week training period (Series A), the animals were rested for 3 weeks and then trained again for 3 consecutive weeks (Series B). The rats in Group 1 were injected daily with saline. The animals in Group 2 were treated daily with 1.5 mg/kg tetrabenazine during the 1st and 2nd week and with saline during the 3rd week.

To illustrate the changes in behavior during the first training period, Table 3.3 shows the performance of the rats on the 1st and 5th day of the 1st and 5th week (Series A). The saline-treated rats (Group 1) developed a stable CAR by the end of the training period. Flash light, the CS, was – with an average of 78.88 ± 11.92% – effective in eliciting the escape of the rats to the other part of the compartment within 10 s. The efficiency of footshock practically reached its maximum. The EFs dropped to an average of 1.90 ± 1.63%. *Treatment with 1 mg/kg tetrabenazine (Group 2) inhibited the performance of the rats significantly.*

As was discussed above, to have a better chance to see whether rats are capable of learning when the catecholaminergic system in the mesencephalon is blocked by tetrabenazine treatment, we were compelled to work in this series of experiment with a strain of low performing ("dull") rats. Especially the females of this strain of rats need a few weeks to fix a stable CAR. Table 3.3 shows that on the 1st day of training only 10.50 ± 4.69% of the conditioned stimulations were effective and by the end of the 1st week only an average of 48.88 ± 9.30% was reached. Under the same training conditions, the rats of a high performing ("clever") strain are capable of escaping in response to flash light on the 1st day of training in more than 80% of the trials (as an example see Knoll et al. 1996b). Table 3.3 also shows that females in this "dull" strain of rats needed a 5-week training period to reach the level of performance that "clever" rats already produce on the 1st day of training.

The saline-treated rats (Group 1) fixed a stable CAR during the 1st training period. In the 2nd training period (Series B), that started after 3 weeks of rest, the CS elicited the escape of rats on the 1st day of training at an average of 82.85 ± 1.86%, and the EFs in response to electric shock reached the zero level (0.95 ± 0.38%).

In Group 2 we raised the dose of tetrabenazine to 1.5 mg/kg after the 3-week resting period, and we treated rats for 2 weeks with this high dose. This treatment blocked almost completely the rats' ability to respond to outside stimuli. On the 5th (last) day of the 2nd week of treatment, there was no sign of conditioned avoidance (average of CARs: 3.50 ± 2.28%) and the average percentage of EFs mounted to 63.13 ± 14.31. Before starting the 3rd week of training on Monday, the rats had their usual rest for two days (Saturday/Sunday). This was sufficient time for the elimination of tetrabenazine and the refilling of the catecholamine stores in the mesencephalic neurons.

During the 3rd week of training the rats of Group 2 were treated with saline. The performance of the rats on the 1st and 5th days are shown in Table 3.3. *The rats treated with tetrabenazine for 7 weeks and trained, produced the same average of CARs (80.75 ± 10.29%) as their saline-treated peers (80.79 ± 2.05%).*

3.4 Cortical Enhancer Regulation

Table 3.3. The modification of behavior of tetrabenazine-treated rats through training in the shuttle box

	Week of training	Performance on the	Group 1. (Control) saline treatment for 5 consecutive weeks		Group 2. Tetrabenazine treatment (1 mg/kg) for 5 consecutive weeks	
			CAR (%)	EF (%)	CAR (%)	EF (%)
Series A	1st	1st day	10.50 ± 4.69	22.50 ± 7.55	5.00 ± 2.76	65.00 ± 12.29*
		5th day	48.88 ± 9.30	3.00 ± 2.37	27.63 ± 12.12	49.00 ± 14.92**
	5th	1st day	63.79 ± 13.80	1.55 ± 1.33	20.00 ± 9.59*	56.38 ± 15.56**
		5th day	78.88 ± 11.92	1.90 ± 1.63	29.00 ± 11.61*	46.38 ± 16.92**
3-week stop in training					Tetrabenazine treatment (1.5 mg/kg) for 2 consecutive weeks	
Series B	1st	1st day	82.85 ± 1.86	0.95 ± 0.38	8.25 ± 5.76***	56.50 ± 12.46**
		5th day	83.40 ± 1.68	0.98 ± 0.42	1.88 ± 0.99***	66.00 ± 13.04**
	2nd	1st day	83.00 ± 1.52	1.00 ± 0.58	1.50 ± 0.85***	65.88 ± 12.89**
		5th day	82.05 ± 1.82	0.95 ± 0.60	3.50 ± 2.28***	63.13 ± 14.31**
					Saline treatment	
	3rd	1st day	80.79 ± 2.05	1.20 ± 0.63	80.75 ± 10.29	10.00 ± 9.29
		5th day	84.50 ± 1.75	1.15 ± 0.53	86.25 ± 1.73	0.75 ± 0.41

Two groups of female rats ($n = 8$) stemming from a strain with very low innate learning capacity were treated in the shuttle box at 100 trials daily for 5 days weekly. The rats were treated subcutaneously 60 min before training with saline and tetrabenazine, respectively. Detailed explanation in the text. *CAR* conditioned avoidance response; *EF* escape failure. Significance in the performance between the saline-, and tetrabenazine-treated rats was evaluated by multi-factor analysis of variance (ANOVA): *$P < 0.05$, **$P < 0.01$, ***$P < 0.001$.

On the 5th day of the 3rd week of training in Series B, there was no difference even in the EFs between the two groups. It is obvious that rats treated with tetrabenazine learned similarly to the saline-treated rats, we were just unable to detect this modification of behavior because mesencephalic enhancer regulation was blocked in Group 2 by tetrabenazine treatment, and the rats were unable to operate in the shuttle box.

The results are in accord with the concept that learning needs only the concurrent operation of functionally different groups of cortical neurons under proper conditions. As the circumstances for the cortical neurons were the same from the beginning until the end in the saline-, and tetrabenazine-treated groups of rats, the modification of behavior was also the same.

The data are in harmony with the working hypothesis that each functionally homogeneous group of cortical neurons is continuously producing its own highly specific enhancer substance, the amount of which is significantly increased when the neuron is activated via its specific stimulus. Considering that the most potent synthetic mesencephalic enhancer substance, (−)-BPAP, exerts its enhancer effect in a range of $10^{-16}-10^{-14}$ M concentration (Knoll et al. 1999), it is reasonable to assume that the cortical enhancer substances also work in a very low concentration. This may throw unusual technical difficulties in the way of detecting and identifying the key actors in cortical enhancer regulation.

3.5
Therapeutic Aspects of Synthetic Mesencephalic Enhancer Substances

3.5.1
The Physiological Mechanisms that Give Reason for the Prophylactic Administration of a Synthetic Mesencephalic Enhancer Substance to Slow Brain Aging

An almost sateless agility, playfulness, and exuberant high spirits after weaning that is extinguished when sexual maturity is reached and a slowly developing decline of behavioral performances during the downhill period of life characterize the brain work of mammalian species capable of acquiring drives.

Our studies on rats suggest that age-related changes in mesencephalic enhancer regulation are primarily responsible

1. For the youthful power of the mammals from weaning until sexual maturity
2. For the transition from the uphill period of life into postdevelopmental longevity
3. For the progressive decay of behavioral performances during the downhill period of life, and
4. For the transition from life to death

3.5 Therapeutic Aspects of Synthetic Mesencephalic Enhancer Substances

Hundreds of millions of people now die at ages over 80, primarily due to the twentieth century progress in hygiene, chemotherapy, and immunology. With a longer average life span the need to improve quality of life during the latter decades of life is more compelling.

It is not to be questioned that brain aging, the unavoidable age-related decay of brain performances, gives a lot of trouble to the aged, causes hardly tolerable inconveniences, and too often makes the latter decades of life a burden. It is therefore hard to overestimate the significance of a safe and efficient prophylactic treatment that slows the aging of the brain.

*

The belief that a steadily acting toxic agent, oxygen, is to blame for age-related deterioration of systems with aerobic metabolism is as simple as it is attractive. Oxygen is a Janus-faced substance: on the one hand, essential for living; on the other, toxic in nature. The cells which use oxygen to maintain their organization and viability use their scavenger systems to fight incessantly against toxic free radicals generated from oxygen. The idea that the life of living beings with aerobic metabolism is ultimately terminated by chronic oxygen toxicity seems to be, at first sight, self-evident and is substantially supported by the shorter life span of species with a higher metabolic rate.

It is common knowledge, however, that cells of vital organs, including the brain, maintain vigorous activity at natural death. The unavoidable chronic oxygen toxicity helps us to understand the progressive, age-related decay of organ function, but cannot explain why natural death sets in exactly at the time it does. *The riddle to be solved is: why does the organism as a whole die, when the aged organs remain fit for life at natural death, even though the passage of time causes deterioration of parts of the system?*

As the brain alone ensures that the mammalian organism works as a purposeful, motivated, goal-directed entity, without denying the significance of the adverse consequences of natural aging in different organs, we may assume that none of them can compete in importance with age-related changes in the CNS.

There are good reasons to assume that it is the physiological role of the catecholaminergic neurons to keep the higher brain centers in a continuously active state, the intensity of which is dynamically changed within broad limits according to need. Such regulation is the condition *sine qua non* for the integrative work of the CNS. The operation of the catecholaminergic system is comparable to an engine which is ignited once for an entire lifetime, as signaled by the appearance of an EEG, in an early phase of development.

Due to aging, the maximum level of activation of the CNS, via the catecholaminergic system, decreases progressively with the passing of time. The blackout ("natural death") of the integrative work of the CNS, signaled by the disappearance of an EEG, occurs when the catecholaminergic system's ability to activate the higher brain centers sinks below a critical threshold and an emergency incident transpires, where a high level of activation is needed to survive and the CNS can no longer be activated to the required extent. This would explain why a common infection, a broken leg, or any other challenge easily surmountable given catecholaminergic machinery working at full capacity may cause death in old age.

The essence of this hypothesis is depicted in Fig. 3.14. According to this scheme, the life of a mammalian organism can be divided, from a functional point of view, into six stages, each beginning with a qualitative change of crucial importance. The first stage starts with the fertilization of the ovum and lasts until the catecholaminergic system properly activates the higher levels of the brain, which then take the lead and integrate the different parts of the organism into a highly sophisticated entity. We may deem the first stage of development of the mammalian organism as completed when the catecholaminergic engine of the brain is put into gear once and for all. This is the intrauterine birth of the unique individual. The appearance of the EEG signals the transition from the first to the second stage of development.

Cells need oxygen, water, and food for life. These are first supplied, via the placenta, by the mother. The subsequent, highly complicated evolving program is devoted to ensuring independence from the mother.

The second stage of development ends with the passage of the fetus from the uterus to the outside world. From a functional point of view birth means the transition from fetal to postnatal circulation, with the newborn infant now supplying itself with oxygen.

The third stage lasts from birth until weaning and serves to develop the skills needed for the maintenance of integrity and for the infant to supply itself with water and food.

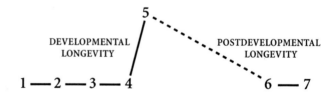

Fig. 3.14. Conception of essential changes during the lifetime of mammals

1) FUSION OF THE SPERMATOZOON WITH THE OVUM
2) THE INTEGRATIVE WORK OF THE CNS SETS IN. APPEARANCE OF EEG
3) BIRTH OF THE FETUS
4) WEANING
5) SEXUAL MATURITY IS REACHED
6) THE INTEGRATIVE WORK OF THE CNS BLACKS OUT. DISAPPEARANCE OF EEG. "NATURAL DEATH"
7) DEATH OF THE LAST CELL

The fourth stage lasts from weaning until, the goal of goals in nature, full-scale sexual maturity is reached. This is the most delightful phase of life, the glorious uphill journey. The individual progressively takes possession, on a mature level, of all abilities crucial for survival and maintenance of the species. It learns to avoid dangerous situations, masters the techniques for obtaining its food, develops procreative powers for sexual reproduction and copulates. This is, at the same time, the climax of developmental longevity.

The sexually fully mature individual fulfils its duty. Thus, to maintain the precisely balanced natural equilibrium among living organisms, the biologically "useless" individual has to be eliminated. According to the inborn program, the fifth, postdevelopmental stage of life (aging) begins.

The essence of the fifth stage is progressive decay of the efficiency of the catecholaminergic system during the postdevelopmental life span until at some point, in an emergency situation, the integration of the parts in a highly sophisticated entity can no longer be maintained and "natural death", signaled by the disappearance of an EEG signal, sets in.

As the parts of the organism remain alive, the sixth and last stage of life is the successive dying off of the different groups of cells.

The hypothesis outlined suggests that the quality and duration of life rests upon the inborn efficiency of the catecholaminergic brain machinery, i.e., a high-performing longer-living individual has a more active, more slowly deteriorating catecholaminergic system than its low-performing, shorter-living peer. To simplify the concept, we may say that a better brain engine allows better performance and a longer life span. *The concept clearly predicts that, as the activity of the catecholaminergic system can be improved at any time during life, it must essentially be feasible to develop a technique for transforming a lower-performing, shorter-living individual into a better-performing, longer-living one. It therefore follows that a shift of the duration of life beyond the technical life span (TLS), with a yet unpredictable upper limit, must be possible in all mammals, including the human species.*

Various species live together on earth in a harmonious proportion. This is obviously carefully regulated. One of the seemingly principal regulatory mechanisms that produces equilibrium among living organisms is brain aging. It ultimately leads to the elimination of those individuals who have already fulfilled their duty of nurturing the new generation.

Now we have to realize that the uphill period of life is epitomized by the operation of the enhancer regulation that maintains the basic activity of the brain on a significantly higher level (Knoll and Miklya 1995). The period of enhanced activity lasts until sex hormones appear, dampen enhancer regulation, and lower the basic brain activity to its preweaning level (Knoll et al. 2000). Thus, sex hormones provide for the transition from the developmental phase of life to postdevelopmental longevity, the period of the slow age-related decay of brain performance and terminated by natural death.

Although the slow and continuous age-related decline of enhancer regulation (the *vis vitalis*) that is characteristic of the downhill, postdevelopmental phase of life starts with the full-scale development of sexual-hormone regulation, it does not mean that the sexually mature individual is immediately converted to a significantly lower performer in its fight for existence. As it was shown earlier in detail, conditioning (learning) makes the performance of the experienced organism highly economic and efficient, even at a lower level of specific activation of the brain (Knoll 1969). Nevertheless, the irresistible, progressive age-related decay of enhancer regulation gradually weakens the compensatory role of experience, and even the most experienced aged organism becomes more and more vulnerable with the passing of time.

Considering the key role of the mesencephalic catecholaminergic neurons in the age-related deterioration of behavioral performances, it stands to reason that to fight against this unavoidable physiological process, there is a need to start a specific prophylactic therapy against brain aging as soon as sexual maturity has been reached. It seems logical that the best chance to realize this aim is the daily small dose administration of a synthetic mesencephalic enhancer substance during the downhill period of life. To justify this conclusion the brain mechanisms that constitute the basis of a prophylactic antiaging therapy need to be surveyed in more detail.

3.5.1.1
Developmental Longevity and Its Termination by Sex Hormones

How can youth be defined; how long does it last; how is it terminated? Or, using a more scientific sounding terminology: *what is the essential difference between developmental and postdevelopmental longevity; what is the cause of the transition from one phase to the other?*

To answer these questions we need to consider a phenomenon of which we first took notice in the course of our behavioral studies on rats performed in the 1950s. We observed that hunger drive induced orienting-searching reflex activity was significantly more pronounced in young rats then in their elder peers (Knoll 1957). We repeatedly corroborated this observation later and described it for the last time in 1995 (Knoll and Miklya 1995.)

Catecholaminergic neurons have a powerful activating effect on the brain. We measured hunger-induced orienting-searching reflex activity in rats and found that animals in the late developmental phase of life (2 months of age) were much more active than those in the early postdevelopmental phase (4 months of age), pointing to enhanced catecholaminergic activity during the developmental phase.

Figure 3.15 shows the striking difference in the intensity of orienting-searching reflex activity of hungry rats in surroundings quite new to them as a function of time elapsed from last feed. Rats in their uphill period of life

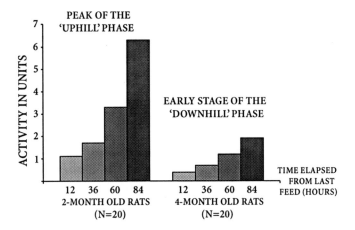

Fig. 3.15. Intensity of orienting-searching reflex activity of hungry rats in surroundings quite new to them as a function of time elapsed from last feed. Activity was measured and expressed in units from 0 to 10. See Knoll and Miklya (1995) for methodology and other details

were much more active than their peers in their early postdevelopmental phase of life.

We have also followed the awakening of sexual drive, maturation of spermatozoa and development of the penis in male CFY rats. In the strain we used in this experiment, it was exceptional to find copulatory drive manifesting in males younger than six weeks. Although the appearance of copulatory patterns usually precedes maturation of spermatozoa and full development of the penis, the overwhelming majority of the males reached full-scale sexual activity by the completion of their 2nd month of life.

3.5.1.1.1
Enhanced Mesencephalic Enhancer Regulation from Weaning Until Sexual Maturity: The Mechanism Responsible for the Exuberant Physical Strength and Mental Vigor in the Uphill Period of Life

In the rat, the interval from weaning (3rd week of life) until the end of the 2nd month of age is decisive for the development of the individual. During this period the animal acquires abilities crucial for survival and maintenance of the species. Based on the observation that 2-month old hungry rats are significantly more active than their 4-month old peers, we checked dopaminergic, noradrenergic and serotonergic activities in the brain before weaning (in 2-week-old rats), during the crucial developmental phase, from weaning to sexual maturity (in 4- and 8-week-old rats), and in the early postdevelopmental phase of life (in 16- and 32-week-old rats). As an indicator of the basic activity of catecholaminergic and serotonergic neurons in the brain, we measured the release of dopamine from the striatum, substantia nigra and tuberculum olfactorium, of norepinephrine from the locus coeruleus, and of serotonin from the raphe, in male and female rats (Knoll and Miklya 1995).

We found that from weaning until the 2nd month of life the striatal dopaminergic system of the rats was significantly more active than either before or after that period. This explains why, as demonstrated in Fig. 3.15, hungry rats in their developmental phase of life were significantly more mobile in an open field than their peers in their postdevelopmental phase of life. Similar age-related changes were detected in the mesolimbic dopaminergic, noradrenergic, and serotonergic systems, too. As an example, Fig. 3.16 illustrates, in male rats, the significantly enhanced release of catecholamines from the mesencephalic catecholaminergic neurons after weaning. Fig. 3.17 shows, in male rats, the dampening of the enhanced mesencephalic enhancer regulation after sexual maturity was reached.

The amount of serotonin released in 4-week-old rats was 7-fold higher in males and 6-fold higher in females than in 2-week-old animals. Thereafter, the amount of serotonin released from the raphe decreased with time, but remained significantly higher than in 2-week-old rats. In contrast, the amount of dopamine released from the striatum or tuberculum olfactorium of 16- to 32-week-old rats did not differ from that measured in 2-week-old animals. The amount of norepinephrine released from the locus coeruleus of 16- to 32-week-old rats was significantly lower than that in 2-week-old animals.

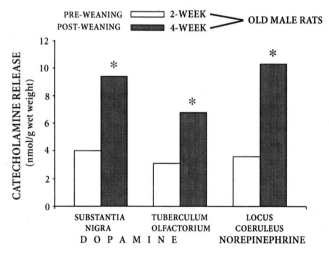

Fig. 3.16. Proof of enhanced catecholaminergic activity in the brain stem of sexually immature rats after weaning. $*P < 0.001$. See Knoll and Miklya (1995) for methodological details

3.5 Therapeutic Aspects of Synthetic Mesencephalic Enhancer Substances

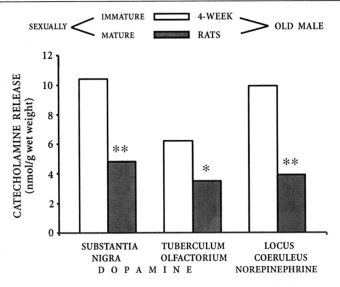

Fig. 3.17. Proof of dampened catecholaminergic activity in the brain stem of rats following sexual maturity. $*P < 0.01$; $**P < 0.001$. See Knoll and Miklya (1995) for methodological details

All in all, we found that enhancer regulation starts working on a higher activity level after weaning, and this state of enhanced activity continues in existence until the completion of full scale sexual development, with a rapid rate of decay thereafter. It is obvious that as soon as sexual maturity was reached the catecholaminergic tone changes from a "hyperactive" to an "economy" state, signaling the transition from a developmental to a postdevelopmental (aging) phase of life, and we may also conclude that enhanced enhancer regulation between weaning and sexual maturity is responsible for the exuberant physical strength and mental vigor of mammals in their uphill period of life.

3.5.1.1.2
Sex Hormones Bring Back the Enhanced Mesencephalic Enhancer Regulation to the Preweaning Level: The Mechanism that Terminates Developmental Longevity and Constitutes the Foundation of the Transition From Adolescence to Adulthood

As it was shown earlier (Knoll and Miklya 1995), we measured in both male and female rats a significantly more pronounced enhancer regulation in the dopaminergic, noradrenergic and serotonergic neurons from the discontinuation of breast feeding (end of the 3rd week of age) until the appearance of sex hormones (end of the 2nd month of life).

Examples of these characteristic changes are shown in Figs. 3.16 and 3.17. Sexual hormonal regulation starts working in the rat with full capacity only at the end of the 2nd month of age. The rapid decrease of the release of norepinephrine, dopamine and serotonin from selected discrete brain regions appeared synchronously with the completion of sexual maturity. Thus it was reasonable to assume that sex hormones dampen the enhancer regulation in the brain stem, and this is the mechanism which terminates developmental longevity.

We castrated three-week-old male and female rats and measured the release of catecholamines and serotonin from selected discrete brain regions at the end of the third month of their life (Knoll et al. 2000). Figure 3.18 shows, for example, that in male rats the amount of catecholamines and serotonin released from the neurons was significantly higher in castrated than in untreated or sham operated rats, signaling that sex hormones inhibit enhancer regulation in the brain.

To further analyze this effect of sex hormones, we treated male and female rats s.c. with oil (0.1 ml/rat), testosterone, (0.1 mg/rat), estrone (0.01 mg/rat) and progesterone (0.5 mg/rat), respectively, and measured their effect on enhancer regulation. Twenty-four hours after a single injection with the hormones, the release of norepinephrine, dopamine and serotonin was significantly inhibited in the testosterone-, or estrone-treated rats, but remained unchanged after progesterone treatment. In rats treated with a single hormone injection, testosterone in the male and estrone in the female was the significantly more effective inhibitor. Remarkably, the reverse order of potency was found in

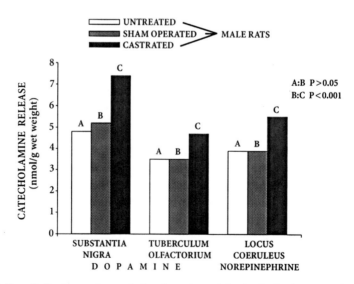

Fig. 3.18. Proof of enhanced catecholaminergic activity in the brain stem of 3-month-old rats castrated at the age of 3 weeks. See Knoll et al. (2000) for methodological details

3.5 Therapeutic Aspects of Synthetic Mesencephalic Enhancer Substances

rats treated with daily hormone injections for 7 or 14 days. After two-week treatment with the hormones, estrone was in the male and testosterone in the female the significantly more potent inhibitor of the enhancer regulation. Figure 3.19 shows, as an example, the dampening of enhancer regulation in male rats treated for 2 weeks with sex hormones.

The data prove that sex hormones terminate the hyperactive phase of life by dampening enhancer regulation in the catecholaminergic and serotonergic neurons. They bring about the transition from the developmental phase of life to postdevelopmental longevity, from adolescence to adulthood. This change is in the meantime also the beginning of the slow, continuous decay of mesencephalic enhancer regulation. As a consequence of it, the fixation of ICRs and the acquisition of drives are subject to an irresistible, slowly progressing, age-related decline until death.

Although individual variation in the age-related decline of behavioral performances is substantial, the process hits every brain. Both the decay in brain performances as well as the potential for the manifestation of age-related neurodegenerative diseases (Parkinson's, Alzheimer's) increases with the physiologically irrepressible aging of the brain. It is obvious that only the development of a safe and efficient prophylactic pharmacological intervention, starting immediately after the completion of sexual maturity, can significantly slow brain aging.

To find, sometime in the future, efficient means to prolong human life beyond the TLS would be a new example of man's endeavor to outwit Nature by understanding the laws of its operation.

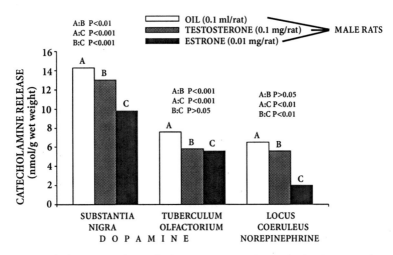

Fig. 3.19. Proof of dampened catecholaminergic activity in the brain stem of 5-week-old rats following a 2-week treatment with sex hormones. See Knoll et al. (2000) for methodological details

3.5.1.2
Postdevelopmental Longevity and its Termination by Natural Death

Aging, the unfortunate common fate of all mature adults, is a physiological phenomenon. It essentially means the decadence of the quality of life with the passage of time. The easily recognizable, external appearance of aging (graying hair, wrinkling skin, use of reading glasses, etc.) gives some information about the chronological age of the person, but these signs are not necessarily in complete harmony with the physiological age of the organ systems, with the measurable decrements of integrated functions (maximum O_2 capacity, maximum breathing capacity, maximum work rate, etc.)or with the almost unmeasurable mental deterioration.

The exact measurement of the age-related changes in man remains difficult because the most reliable technique for following the changes in a given individual over his or her entire life span is practically unfeasible. The available information about the age-related changes in human population stems from cross-sectional studies, from the comparison of differences in performances between different age groups.

The main problem is, however, that the scatter within a particular age group for any measurable parameter is extreme. The reason for this extreme variation is the lack of a general factor of physiological age. In cross-sectional studies no single age emerges as the point of sharp decline in function. Any individual may show different levels of performance and the careful observer finds many dissociations between "chronological" and "physiological" age. A 70-year-old man, for example, may retain sexual performance equal to that of 40-year-olds, but display a cardiac output equivalent to his peers and a visual memory equivalent to average values for 80-year-old subjects. An additional factor which explains the extreme difficulties in measuring age-related changes with statistical methods lies in the nonlinear nature of the regression of a number of functions. Despite all these weaknesses, the average life span in the most developed countries has already exceeded the 80-year level. This change has come about due to the prevention of premature deaths owing to the development of hygiene, immunology and chemotherapy. The technical life span (TLS) of the human race, close to 120 years, has remained, however, unchanged.

*

It was discussed above (Sect. 3.5.1) that

1. Since the brain alone ensures that the mammalian organism works as a purposeful, motivated, goal-directed entity, the age-related changes in the CNS are of particular importance.
2. Since the enhancer-sensitive neurons in the brain stem work as the engine of the brain, the slow, continuous, postdevelopmental functional decline of

mesencephalic enhancer regulation is of primary importance in the maintenance of the well-balanced equilibrium among living organisms, because it helps to eliminate the individuals who already fulfilled their duty in nurturing the new generation.

For the time being the prestigious task – the maintenance of mesencephalic enhancer regulation during the postdevelopmental phase of life on the enhanced level characteristic of developmental longevity – cannot be fully accomplished. However, it is already feasible to modestly slow the age-related decay of the catecholaminergic and trace-aminergic tone in the brain via the prophylactic administration of 1 mg (−)-deprenyl daily.

The development of (−)-BPAP, an at least hundred times more potent synthetic mesencephalic enhancer substance than (−)-deprenyl, is by itself a hint that our present knowledge about mesencephalic enhancer regulation is in a very early stage. The high potency of (−)-BPAP indicates already that much more potent natural enhancer substances than PEA and tryptamine might exist. Better understanding of mesencephalic enhancer regulation promises to develop more efficient techniques in the future to slow brain aging and prolong human life beyond the TLS.

3.5.1.2.1
The Slow Decline of Mesencephalic Enhancer Regulation from Sexual Maturity Until Death: The Mechanism of Brain Aging Primarily Responsible for the Downhill Period of Life

To get an overall picture of the slow, continuous functional decline of the mesencephalic enhancer regulation during the postdevelopmental phase of life we need to follow the age-related decay in the state of supply of the mammalian brain with important endogenous substances related to this regulation. Let us set out on the track of PEA and dopamine.

To date PEA is the most carefully investigated natural enhancer substance (see Sect. 3.1.2). The first note in the literature furnishing indirect evidence that PEA may be an endogenous CNS stimulant in humans was the finding of Fischer et al. (1968). They found that the urinary excretion of free PEA was reduced in depressed patients and suggested that a PEA deficit may be one of the biochemical lesions leading to depression. Experimental evidence was soon presented that PEA is an endogenous constituent of the mammalian brain (Fischer et al. 1972; Saavedra 1974; Wilner et al. 1974). Sabelli and Mosnaim (1974) expounded the hypothesis that PEA might play a physiological role in affective behavior.

Papers discussing the possible role of PEA as a physiological mood elevator (Greenshow 1989; Davis and Boulton 1994; Sabelli and Javaid 1995; Sabelli et al. 1986, 1996; Premont et al. 2001) as well as papers proposing a role of trace amines in a series of illnesses, such as schizophrenia, depression, attention deficit/hyperactive disorder, Parkinson's disease, Rett's syndrome, migraine, phenylketonuria, hepatic encephalopathy, and hypertension (Usdin and Sandler 1976; Boulton et al. 1988; Saavedra 1989; Walker et al. 1996; Janssen et al. 1999; Satoi et al. 2000; Premont et al. 2001) were continually published.

An important step forward in the history of trace amines was the discovery of the presence of high-affinity binding sites for tyramine, tryptamine, and PEA (Hauger et al. 1982; Nguyen and Juorio 1989; Nguyen et al. 1989). As already previously mentioned, the binding sites have been identified as G-protein-coupled trace-amine receptors (see Sect. 3.3); the expression of trace-amine receptor mRNA in human amygdala further suggests a role of trace amines in depression and anxiety disorders (Borowsky et al. 2001); and methamphetamine, the PEA-derivative with a long-lasting effect, is also a trace-amine receptor agonist (Bunzow et al. 2001).

The discovery that PEA is an natural enhancer substance (see Chap. 3) clarified the mechanism of the stimulatory effect of PEA and finally assigned this trace amine its physiological role in the regulation of behavioral performances (see Shimazu and Miklya 2004, for review). Long before the discovery of the enhancer effect of PEA (Knoll et al. 1996c), we had already established that during postdevelopmental longevity there is a continuously increasing PEA deficit in the mammalian brain (Knoll 1982). This paper put forth the thesis that the progressive decrease in brain catecholamines and trace amines is an unavoidable biochemical lesion of aging. This concept was based, on the one hand, on the enhanced MAO-B activity in the aging brain, and on the other hand, on the antiaging effect of (−)-deprenyl, the first highly potent and selective inhibitor of MAO-B described.

As a rule, enzyme functions decrease in the brain with the passing of time. B-type of MAO is an exception. Robinson et al. (1971, 1972) published the first papers demonstrating that MAO activity progressively increases in the aging brain. This finding was corroborated within a couple of years by different groups (Nies et al. 1973; Mantle et al. 1976; Shih et al. 1979; Carlsson 1979; Eckert et al. 1980; Fowler et al. 1980a,b; Strolin Benedetti and Keane 1980). It soon became clear, however, that both in the brain of humans (Fowler et al. 1980b) and rats (Mantle et al. 1976; Strolin Benedetti and Keane 1980) only the activity of the B-type of MAO is increased in the aged. It was also shown that the selective age-dependent decrease in MAO-B activity was due entirely to an increased enzyme concentration in brain tissue (Fowler et al. 1980b).

Student and Edwards (1977) demonstrated that MAO-B is predominantly localized in the neuroglia, a finding soon corroborated (Strolin Benedetti and Keane 1980) and now firmly established as fact.

In my hypothesis, put forth in 1981–1982, I suggested that a progressively developing biochemical lesion in the aging brain which leads to catecholaminergic and trace-aminergic deficiency is responsible for the age-related decline

3.5 Therapeutic Aspects of Synthetic Mesencephalic Enhancer Substances

in sexual and learning performances and ultimately leads to natural death (Knoll 1981a,b,c, 1982).

Let us quote here the original description of the hypothesis (Knoll 1982, pp. 109–110):

> [W]ell established old experiences offer a good explanation for the increase of brain MAO-B activity in the latter decades of life. Cell loss is a general feature of the aging brain ... As the loss of neurons is always compensated by glial cells, the progressive and cumulative loss of neurons in the aging brain gives a satisfactory explanation to the selective increase of extrasynaptosomal MAO-B activity with increasing age. This seems to be an unavoidable biochemical lesion of aging ... Collating the facts that there is an unavoidable loss of neurons, inescapably leading to increased MAO-B activity with increasing age, makes it understandable that dopaminergic and trace-aminergic modulation in the brain is progressively decreasing in the aging brain. It is in agreement with this trend of changes that an age-dependent decrease in the dopamine control of the basal ganglia in man was described, first by Bertler (1961), and corroborated by many others. Riederer and Wuketich (1976) found that the dopamine content of the human caudate nucleus decreased in an age-related manner.
>
> If, in addition, we also consider that the activity of tyrosine hydroxylase, the enzyme catalyzing the rate-limiting step in catecholamine biosynthesis, was also found to decrease in human brain tissue with increasing age (McGeer et al. 1971), weighty arguments seem to support the view that catecholaminergic tone is progressively decreasing in the aging brain.
>
> As the described age-dependent chain of events can be deduced to well-defined biochemical lesions, the chances of developing a new drug strategy for counteracting or possibly even preventing the adverse consequences of the age-related decrease of the catecholaminergic tone in the brain are fair.

Dozens of studies published since the proposal of this hypothesis have strengthened this approach step-by-step. The evolution of the project can be followed via the reviews published after 1982, when the original hypothesis was presented, until the discovery of mesencephalic enhancer regulation (Knoll 1983, 1985, 1986a,b,c, 1989, 1992a,b, 1993a,b,c, 1995, 1998, 2001, 2003).

In the light of our present knowledge there can be little doubt that because of the continuously increasing MAO-B activity in the aging brain, the more and more efficient metabolism of PEA necessarily works against the chances of a freshly synthesized PEA molecule reaching its target. This is one of the factors which contributes to the age-related decline of mesencephalic enhancer regulation with the passing of time.

The same fits for dopamine. Figure 3.20 shows the decay in the dopamine content of the caudate nucleus in the aging human brain. We lose 13% of our brain dopamine in the decade after age 45. At this normal rate nobody will exceed, within the obtainable human life span, the critical threshold of dopamine content (30%) that accompanies the precipitation of the symptoms of Parkinson's disease. Thus, as illustrated in Fig. 3.20, Parkinson's disease is obviously

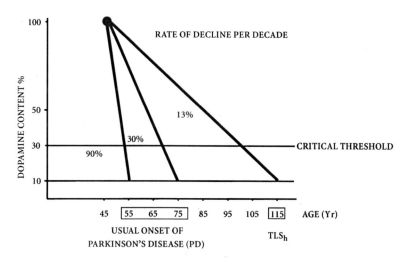

Fig. 3.20. Visualization of the concept that Parkinson's disease (*PD*) is the premature rapid aging of the striatal dopaminergic machinery. TLS_h technical life span in humans

the consequence of a premature aging of the dopaminergic machinery in the human brain. To realize the functional consequences of the age-related decline of the dopaminergic system in the mesencephalon, let us follow, as an example, the decline of the ejaculatory activity in human and rat males during their postdevelopmental phase of life.

Sexual activity in the human male is known to be influenced by a number of factors, such as good health, stable marriage, satisfactory sexual partner(s), and adequate financial and social status. But even in the males who meet all the requirements for retention and maintenance of sexual functioning, there is an age-related decrease in sexual vigor.

Martin (1977) studied coital activity as a function of age, interviewing 628 members of the Baltimore Longitudinal Study of Aging. The subjects were white, married, urban residents in good health from the Washington-Baltimore area, varying from 20 to 95 years of age. According to this study, median coital activity was highest, 2.1 events/week, between the ages of 30 and 34, and decreased progressively with increasing age, sinking to 0.2/week in the age group 65–69.

The study throws light upon the enormous individual variation in sexual vigor. The mean frequency of total sexual activity in 159 males was found to be 520 sexual events/5 years in the age-group 20–39, including young males performing below 100 sexual events/5 years and those with frequencies of total sexual activity over 1,000 sexual events/5 years.

3.5 Therapeutic Aspects of Synthetic Mesencephalic Enhancer Substances

In the age group 65–79, the mean frequency of total sexual activity decreased to 75 sexual events/5 years, but even in this group subjects producing 400–700 sexual events/5 years were registered.

Table 3.4 summarizes the data of the human study to illustrate the age-related decline of coital activity and the striking individual differences in the sexual performance of human males in different age cohorts.

In a number of longitudinal studies performed on male rats we observed that the age-related decline of coital activity in male rats and the striking individual differences in sexual performance in different age cohorts are essentially the same as in human males. Table 3.5 shows the age-related decline of coital activity in male rats and the striking individual differences in copulatory activity in different age cohorts.

Because of brain aging, even the sexually high performing males necessarily lose their potency to ejaculate if they live long enough. In our studies on male CFY rats we followed the sexual performance of the animals once a week from sexual maturity until death. We measured three patterns: mounting, intromission, and ejaculation. We found that in response to brain aging even the best performing individuals lost their potency to ejaculate not later than at the completion of their second year of age (see Table 3.5).

We published a series of experiments in 1988 (see Table 3 in Knoll 1988), the results of which clearly show in retrospect that the age-related functional decline of the sexual performance of male rats can be taken as an indicator of the decay of mesencephalic enhancer regulation with the passing of time. We selected 132 sexually inexperienced 24-month-old rats for this experiment and tested their sexual activity in four consecutive weekly mating tests. Table 3.5 shows that none of them displayed ejaculation during the test period. Out of 132 animals, 46 did not show any sign of sexual activity, 42 displayed mountings only, and 44 displayed mountings and intromissions (sluggish rats).

Table 3.4. The age-related decline of coital activity in human males and percentages of striking individual differences from average copulatory activity in different age cohorts. Based on data from Martin (1977)

Age category	Age (years)	Average performance (sexual events/5 years)	Percentage of striking differences in performance from the average
Young adult males	20–39 (30–34)	~ 500 (~ 600)	~ 4% below 100 ~ 25% over 1,000
Mature adult males	40–49	300	~ 8% below 100 ~ 15% over 400 ~ 8% over 600
Aging and aged	65–80	Below 100	~ 3% over 400 ~ 1.5% over 600

Table 3.5. The age-related decline of coital activity in male CFY rats and the striking individual differences in the copulatory activity in different age cohorts

Age of rats (months)	Number of rats (n)	Complete inactivity n (%)	Mountings only n (%)	Mountings and intromissions n (%)	Full-scale activity (ejaculations) n (%)
3–6	381	21 (5.51)	20 (5.24)	140 (36.74)	200 (52.49)
12–18	138	27 (19.56)	27 (19.56)	76 (55.07)	8 (5.80)
24	132	46 (34.84)	42 (31.82)	44 (33.33)	0

Of the 132 rats, we treated half of the population (66 rats) with saline and we followed the sexual performance of the animals once a week until they died. The results shown in Table 3.6 allow the conclusion that the duration of the life of the rats was inversely proportional to their sexual performance. As sexual performance is directly proportional to the functional state of enhancer regulation in the dopaminergic neurons, we assume that the rat dies when the age-related decline in mesencephalic enhancer regulation arrives at a critical threshold. The finding that regarding sexual performance: Group 1 < Group 2 < Group 3 is just a sign that the rats belonging to Group 1 are the closest to exceeding the critical threshold resulting in natural death and die out first, rats in Group 2 live longer, and rats in Group 3 are the longest living.

The age-related decline in mesencephalic enhancer regulation during the postdevelopmental phase of life in male rats can further clearly be recognized by comparing the individual variation in sexual performance of 3- to 6-month-old male rats with the performance of their 2-year-old peers. Table 3.5 shows that whereas 52.49% of 3- to 6-month-old male rats displayed ejaculations during the four consecutive mating tests, only 5.80% of 12- to 18-month-old males ejaculated, and none of the 24-month-old males was in possession of this faculty any longer.

Moreover, the age-related change in the percentage of animals belonging to the "noncopulator" group clearly proved that mesencephalic enhancer regulation is in continuous decline during the postdevelopmental phase of life. Only 5.51% of the 3- to 6-month-old males were sexually inactive, but 19.56% of the 12- to 18-month-old rats and 34.84% of the 24-month-old rats belonged to this group.

3.5.1.2.2
The Relative Weakness of the Brain Engine at the Time of an Emergency Incident (Natural Death): The Mechanism that Terminates Postdevelopmental Longevity

The rapid, age-related decline of the activity of the mesencephalic catecholaminergic system, which plays a key role in the activation of the cortex, suggested a reasonable explanation for the onset of natural death (see Knoll

3.5 Therapeutic Aspects of Synthetic Mesencephalic Enhancer Substances

Table 3.6. The dying out of 66 saline-treated male rats with differing sexual activities. Data taken from Knoll (1988)

Week of saline treatment	Number of dying animals in		
	Group 1 (noncopulators) ($n = 23$)	Group 2 (mountings only) ($n = 21$)	Group 3 (sluggish rats) ($n = 22$)
36th	2		
37th	4		
38th	8	1	
39th	-	-	
40th	3	-	
41st	5	1	
42nd	1	7	
43rd		6	1
44th		3	4
45th		-	1
46th		2	2
47th		1	5
48th			3
49th			1
50th			2
51st			-
52nd			-
53rd			-
54th			-
55th			-
56th			2
57th			-
58th			-
59th			-
60th			1

1994, for review). According to this working hypothesis postdevelopmental longevity is terminated because the functional deterioration of the mesencephalic catecholaminergic machinery that activates the cortex sinks below a critical threshold in the aging brain. At the time of an emergency incident, when a high level of activation is needed to survive the tribulation, the cortical neurons cannot be further activated to reach the required level. Thus the essence of the natural death situation is the *relative* weakness of the brain engine. This would explain why a common infection, a broken leg, or any other challenge easily surmountable given catecholaminergic machinery working at full capacity may cause death in old age. Continuous age-related decline of the brain engine's performance makes it unavoidable that with the passing of time every otherwise healthy living being arrives at natural death.

Due to inborn differences in the efficiency of the brain engine, we have in a random rat population extreme differences in performance. We can therefore create experimental circumstances in low-performing young rats leading to the *relative* weakness of the brain engine. We thus mimic a natural death situation to which otherwise healthy animals full of vigor fall victim.

In our effort to find for the rat the most favorable conditions for the acquisition of the glass-cylinder-seeking drive, we analyzed the kinetics of escape from a hot plate at different temperatures (see Chap. 3, pp. 43–53, in Knoll 1969, for review). The essence of the method (see Fig. 3 in Knoll 1969) was a copper hot plate (180 × 180 mm) the temperature of which could be regulated with an accuracy of ±0.2 °C. The metal surface was heated to the desired temperature and a glass cylinder, 30 cm high, with bottom and top diameters of 16 and 12 cm, was placed at the top of it. The glass cylinder, opened on bottom and top, had no side opening. The experiment was performed on female rats weighing 120–150 g. The animals were dropped onto the heated metal plate, through the upper opening of the glass cylinder. The time elapsing between their fall and subsequent jump onto the cylinder's top was measured. Ten such tests were usually performed at 30 s intervals. If the animal had not jumped out of the glass cylinder in 4 h in any of the tests, no further measurements were made and the 4 h time was referred to as the maximum value.

Although the experimental circumstances are peculiar, in essence we activate an inner drive. The animal fights for survival and mobilizes all resources to find the way of escape by trial and error. What definitely happens in the brain is that the heat/pain stimulation activates the brain engine and in proportion with it the cortical neurons start working on a high-activity level, allowing the acquisition of those chains of ECRs that ultimately enables the rats to find, by trial and error, the way to escape from the hot plate. Rats with inadequately activated brain engines are in danger.

At 40 °C, the animals stayed in the glass cylinder throughout the whole period of 240 min, thus, this stimulus was subliminal to properly activate the brain engine and cortical neurons toward eliciting the escape reaction. A temperature of 45 °C was the lowest that elicited jumping onto the rim of the upper opening of the glass cylinder. The time needed for escape at this temperature varied between 25 and 164 min in a group of naive rats ($n = 10$) (see Table 13 in Knoll 1969). These conditions were harmless and the experiment did not change the health of any of the rats.

Raising the temperature of the hot plate to 50 °C changed the situation significantly. This not too intensive heat/pain stimulus is insidious and becomes

3.5 Therapeutic Aspects of Synthetic Mesencephalic Enhancer Substances

life threatening for a low-performing rat. At this temperature six out of a group of ten naive rats escaped within 2–13 min and remained healthy. Four rats, however, were unable to cope with the situation; three of them died and the fourth became seriously ill. Table 3.7 shows the corresponding details.

Heating the hot plate to 55 °C yielded a more favorable result. Nine out of a group of ten rats escaped between 1 and 6 min and remained healthy. One animal needed 14 min to escape and later became seriously ill (see Table 15 in Knoll 1969).

The heat/pain stimulation via a hot plate at 60 °C confused some of the animals and, out of a group of ten, three died and one became ill (see Table 16 in Knoll 1969). Nevertheless, for producing glass-cylinder-seeking rats we used this temperature as it proved to be the most efficient.

The success of the escape of naive rats from the hot plate depends primarily on the efficiency of their mesencephalic catecholaminergic machinery. This was shown in rats pretreated for 4 weeks, once daily, with 0.1 mg/kg reserpine. None of a group of ten reserpine-treated rats escaped from a hot plate heated to 55 °C; they all died. However, seven out of ten reserpine-treated rats safely escaped when the temperature of the hot plate was raised to 60 °C (see Table 19 in Knoll 1969).

Altogether, the data substantially support the view that natural death sets in when the catecholaminergic system's ability to activate the telencephalon sinks, for whatever reason, below a critical threshold, and during a state of emergency when a high level of activation is needed to survive the tribulation, the cortical neurons cannot reach the necessary level of activation.

The results shown in Table 3.7 are especially apt in support of this conclusion. There are innate differences in the efficiency of the mesencephalic catecholaminergic system. For the brain of animals (nos.) 1, 2, 4, 5, and 6 the hot plate heated to 50 °C was already an adequate stimulus to enhance the excitability of the cortical neurons for escape and to survive the tribulation without perceivable consequences. Animals 3 and 7 escaped, but fell ill thereafter. Animals 8, 9, and 10 were unable to cope with the situation and died.

3.5.2
Rationale for Slowing the Age-Related Decline of Mesencephalic Enhancer Regulation by the Daily Administration of a Small Dose of a Synthetic Mesencephalic Enhancer Substance from Sexual Maturity Until Death

In light of the peculiar changes in enhancer regulation during the developmental phase of life, the antiaging potential of the administration of a small dose of a safe enhancer substance during the postdevelopmental (aging) phase of life deserves serious consideration. It seems reasonable to shift safely the functional constellation of the brain during postdevelopmental longevity to-

Table 3.7. Escape of rats from a hot plate at 50 °C. Optimal temperature to mimic a "natural death" situation for low-performing rats

Serial no.	Jumping reaction times in rounded-off minutes on experiment day										Note
	Day 1	Day 2	Day 3	Day 4	Day 5	Day 6	Day 7	Day 8	Day 9	Day 10	
1	2	1	2	1	1	1	1	1	1	1	
2	2	1	2	2	2	1	1	1	1	1	
3	34	1	2	3	12	11	3	3	1	1	ill
4	10	1	1	4	7	1	1	1	1	1	
5	13	7	4	8	5	7	2	1	1	1	
6	4	1	3	3	13	1	1	1	1	1	
7	8	7	4	2	2	8	2	1	2	2	ill
8	34	14	32	2	22	6	6	2	–	–	ill, died
9	49	38	–	–	–	–	–	–	–	–	ill, died
10	63	–	–	–	–	–	–	–	–	–	ill, died
Mean	22	8	6	3	8	5	2	1	1	1	

Mean rectal temperature on day 1: immediately after 1st measurement = 38.8 ± 0.9 °C, immediately after 10th measurement = 39.7 ± 1.0 °C, 30 min after 10th measurement = 37.1 ± 0.4 °C
See Chap. 3 in Knoll (1969) for methodological details

3.5 Therapeutic Aspects of Synthetic Mesencephalic Enhancer Substances

wards this one characteristic of the uphill period of life (from weaning until sexual maturity).

To keep the engine of their brain on a higher activity level during postdevelopmental longevity, humans need lifelong medication. It is reasonable to start with the daily administration of a small amount of a synthetic mesencephalic enhancer substance immediately following sexual maturity. This will work for decades. It will improve the quality of life in the latter decades, increase the chances for a longer life, and decrease the danger of precipitating age-related depression and neurodegenerative diseases (Parkinson's, Alzheimer's).

As to the mechanism of the antiaging effect of synthetic mesencephalic enhancer substances, we need to recapitulate the essence of the enhancer effect as was shown in Figs. 3.1 and 3.2. Due to enhanced excitability, a higher number of the enhancer-sensitive neurons in the brain stem get activated to a given stimulus under the influence of natural enhancer substances, PEA and tryptamine. Synthetic mesencephalic enhancer substances, (−)-deprenyl and (−)-BPAP, act similarly (see Sect. 3.1.3). Due to the age-related continuous, slow decline of mesencephalic enhancer regulation during postdevelopmental longevity, the number of the enhancer-sensitive mesencephalic neurons which are activated in response to a given stimulus decreases proportionally over time. Since synthetic mesencephalic enhancer substances increase the excitability of the enhancer-sensitive mesencephalic neurons, it can reasonably be expected that their proper administration during the postdevelopmental phase of life will slow the age-related decay of mesencephalic enhancer regulation. In this sense a synthetic mesencephalic enhancer substance is an antiaging drug.

As an example, the functional decline of the mesencephalic dopaminergic neurons with the passing of time is convincingly demonstrated both in human and rat males in Tables 3.4 and 3.5. The conspicuous age-related decay of sexual performance is the proof of the decline of enhancer regulation in the mesencephalic dopaminergic neurons in aging human and rat males. The same stimulus that activated a high number of the dopaminergic neurons in the young males will activate lower and lower number of these neurons as time passes. Since the administration of a synthetic mesencephalic enhancer substance will increase, by its specific effect, the excitability of the dopaminergic neurons, treatment of an aged male with such a compound can be expected to enhance sexual activity and bring it back to a level that was characteristic of an earlier stage of life of the individual. We have proved experimentally in a series of studies the validity of this assumption.

To date longevity studies have been performed only with (−)-deprenyl, as this compound was the only available synthetic mesencephalic enhancer substance. To illustrate the antiaging effect of (−)-deprenyl, Table 3.8 shows some data taken from our first longevity study on male rats performed between 1984 and 1988. We started the experiments with 2-year-old male rats. Before

Table 3.8. Antiaging effect of (−)-deprenyl treatment. Data taken from Table 4 in Knoll (1989). Details explained in text

Classification of the groups according to sexual performance before treatment	Number of animals	Total number of mountings (M), intromissions (I) and ejaculations (E) of the groups during treatment		
		M	I	E
Saline-treated rats				
Non-copulators	23	37	0	0
Mounting rats	21	425	54	0
Sluggish rats	22	477	231	0
(−)-Deprenyl-treated rats				
Non-copulators	23	997	544	190
Mounting rats	21	1,129	662	172
Sluggish rats	22	1,696	1,257	481

treatment the sexual performance of the rats was measured in four consecutive weekly mating tests. Three of the characteristic patterns of sexual behavior, mounting (M), intromission (I), and ejaculation (E) were monitored. Rats were classified as "noncopulators" (no sign of sexual activity), "mounting rats" (displayed mountings only), and "sluggish rats" (displayed mountings and intromissions) (Knoll et al. 1983; Dalló et al. 1986a,b; Knoll et al. 1989).

In the four mating tests performed before treatment, out of the 132 males selected for this experiment, 46 were found to be noncopulators; 42 animals displayed mountings only, and 44 rats proved to be sluggish. None of the rats displayed ejaculation. It was our intention to start the experiment with 2-year-old rats as we knew from our earlier studies that the strain of CFY rats we used in our experiments lost the ability to ejaculate before they completed their 2nd year of age.

After classifying our male rats according to their sexual performance in the testing period, we started treating rats s.c. with saline (1 ml/kg) and (−)-deprenyl (0.25 mg/kg), respectively, three times a week, until they died. We tested their sexual performance once a week. In each group half of the animals were treated with saline and half with (−)-deprenyl.

Table 3.8 shows the total number of mountings, intromissions, and ejaculations displayed by the whole group of rats until they died. The noncopulator group of the saline-treated rats displayed only 37 mountings altogether. In harmony with the results of the four initial mating tests, these rats were really sexually inactive. In the saline-treated group of rats that displayed mountings only in the initial test, the total number of mountings (425) was much higher than in the noncopulator group, and altogether 54 intromissions were produced. The saline-treated sluggish rats displayed 251 intromis-

3.5 Therapeutic Aspects of Synthetic Mesencephalic Enhancer Substances

sions. As expected, no single ejaculation was detected in the 66 saline-treated males.

In the (−)-deprenyl-treated group the total number of mountings and intromissions increased tremendously and ejaculations were also displayed. Thus due to the enhancer effect of (−)-deprenyl the excitability of the mesencephalic dopaminergic neurons was increased and the efficiency of sexual performance of the rats changed accordingly.

To further illustrate the peculiar antiaging effect of a synthetic mesencephalic enhancer substance, the results of characteristic human and rat studies are summarized in Figs. 3.21 and 3.22, respectively.

Figure 3.21 illustrates the findings of the DATATOP study performed on selected patients with early, untreated Parkinson's disease (Parkinson Study Group 1989). The 401 placebo-treated patients (P) needed levodopa significantly earlier than the 399 (−)-deprenyl-treated patients. The reason for the observed effect is clear. The diagnosis of Parkinson's disease indicates that the decrease in the dopamine content of the caudate nucleus had already exceeded the critical threshold (30%). Because the dopaminergic system continues to deteriorate, the next precisely measurable step of decay is the point at which the patient requires levodopa. Fig. 3.21 shows that

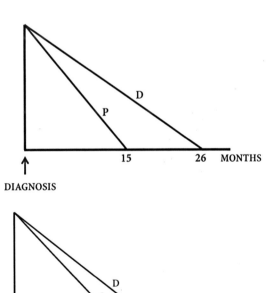

Fig. 3.21. The rate of deterioration from diagnosis to the need for levodopa in parkinsonian patients treated with placebo (P) and (−)-deprenyl (D), respectively. Based on the data published by the Parkinson Study Group (1989)

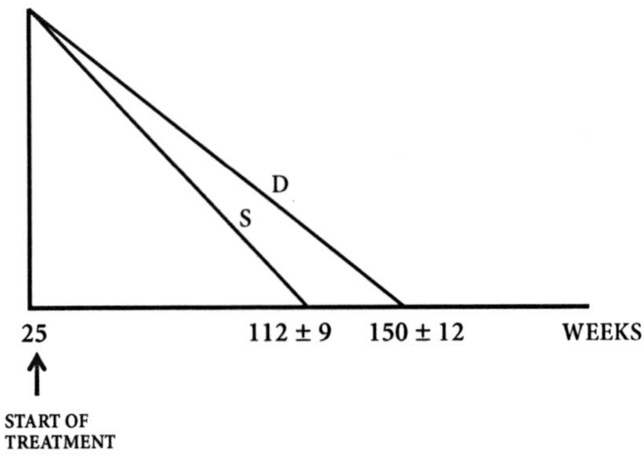

Fig. 3.22. The age-related decline of the sexual potency of male CFY rats treated with saline (S) and (−)-deprenyl (D), respectively, until they lost the ability to ejaculate

the enhancer effect of (−)-deprenyl significantly delayed the onset of this stage.

Figure 3.22 shows essentially the same effect of (−)-deprenyl on male CFY rats. The experiment was performed on 90 selected young males possessing full-scale sexual activity. Half of the population was treated with saline (1 ml/kg), the other half with (−)-deprenyl (0.25 mg/kg), three times a week, from the 25th week of age. The rats' sexual performance was tested once a week. In this study the loss of the ability to ejaculate was selected as the age-related end stage. Saline-treated rats reached this stage at an average of 112 ± 9 weeks. In contrast, (−)-deprenyl-treated rats reached it at an average of 150 ± 12 weeks ($P < 0.001$) (Knoll 1993a). As sexual performance is a dopaminergic function, it is obvious that the enhanced activity of the mesencephalic dopaminergic neurons was responsible for the significantly retarded loss of the ability to ejaculate in the (−)-deprenyl-treated group.

*

Due to its enhancer effect (−)-deprenyl protects the nigrostriatal dopaminergic neurons against selectively acting neurotoxins (Knoll 1978; Hársing et al. 1979; Cohen et al. 1984; Finnegan et al. 1990; Vizuete et al. 1993; Wu et al. 1993), and facilitates scavenger function in the striatum (Knoll 1988, 1990; Carrillo et al. 1991, 1992).

In a series of experiments we also succeeded in finding morphological evidence that treatment of rats with a synthetic mesencephalic enhancer substance slows aging of the neurocytes of the substantia nigra of rats. We developed a method, using a TV-image analyzer, to compare different morphological parameters in the substantia nigra of young and old male rats (Tóth et al. 1992). With the aid of this method, we determined the number, total area, area of one granule, and density features (sum and average of gray values and average gray value of one pigment granule) of melanin granules in neurocytes of the substantia nigra in 3-month-old and 3-year-old male rats (Knoll et al. 1992b). The number of cells in sections of identical areas was similar in the young and old rats. A statistically nonsignificant difference between the two age cohorts in the proportion of neurocytes with and without melanin was found: 773 (48.1%) with and 853 (51.8%) without in the young rats and 1,219 (65.1%) and 652 (34.8%) in the old ones. Within the melanin-containing neurocytes, however, statistically significant age-related differences in the number, area, and density features of melanin granules were discovered. The majority of the neurocytes in young rats contained numerous, small-sized neuromelanin granules, whereas in the majority of the neurocytes of old rats, smaller numbers of large-sized neuromelanin granules were detected.

Table 3.9 shows examples of the characteristic, highly significant differences in the morphological fine structure of the neuromelanin granules in the neurocytes of the substantia nigra of 3-month-old and 3-year-old rats.

3.5 Therapeutic Aspects of Synthetic Mesencephalic Enhancer Substances

The statistically significant age-related differences revealed by the TV-image analyzer allowed us to check morphologically the influence of long-term (−)-deprenyl treatment on the neurocytes of the substantia nigra.

Table 3.10 shows that (−)-deprenyl, injected subcutaneously three times a week for 18 months, prevented the age-related morphological changes of the pigment granules in the neurocytes of the substantia nigra of rats. This is morphological evidence for the antiaging effect of prophylactic treatment with a synthetic mesencephalic enhancer substance.

Rinne et al. (1991) found that the number of medial nigral neurons was greater and the number of Lewy bodies fewer in those parkinsonians who had been treated with (−)-deprenyl in combination with levodopa when compared with patients who had received levodopa alone.

Table 3.9. Age-related differences in the number of neuromelanin granules and in the area of one granule in the neurocytes of the substantia nigra. Data taken from Table 3 in Knoll (1992b)

	Neurocytes from				
	3-month-old rats ($n = 773$)		3-year-old rats ($n = 652$)		
	Mean	SD	Mean	SD	P
Number of neuromelanin granules in a neurocyte	2,600	1,500	1,800	900	< 0.0001
Area of one granule (μm^2)	49,000	18,800	116,300	29,500	< 0.0001

Table 3.10. Age-related changes in the number of neuromelanin granules and in the area of one granule in the neurocytes of the substantia nigra and prevention of these changes by (−)-deprenyl treatment (0.25 mg/kg, s.c. three times a week for 18 months). Data taken from Table 3.3 in Knoll (1992b)

			Neurocytes from			
	3-month-old rats (YC) ($n = 473$)		21-month-old rats treated for 18 months with saline (OC) ($n = 481$)		21-month-old rats treated for 18 months with (−)-deprenyl (DT) ($n = 503$)	
	Mean	SD	Mean	SD	Mean	SD
(1) Number of neuromelanin granules in a neurocyte	2,000	1,100	1,700	920	2,000	1,100
(2) Area of one granule (μm^2)	39,600	25,000	75,000	30,600	39,800	19,600

(1) YC vs OC $P < 0.002$; YC vs DT $P > 0.05$
(2) YC vs OC $P < 0.0001$; YC vs DT $P > 0.05$

3.5.3
Clinical Experiences with (−)-Deprenyl in Depression and in Neurodegenerative Diseases: Further Therapeutic Prospects

Depression. The logic indication for the use of synthetic mesencephalic enhancer substances is their prophylactic administration to slow the physiological age-related decay of mesencephalic enhancer regulation. Nevertheless, as discussed already earlier (see Sect. 3.1.2), (−)-deprenyl, the PEA-derived synthetic mesencephalic enhancer substance, was originally developed with the intention to use it as a new spectrum antidepressant.

The antidepressant effect of the compound was first demonstrated by Varga (1965) and Varga et al. (1967) with the racemic form, and in 1971 with the (−) enantiomer (Tringer et al. 1971). The first study that corroborated the antidepressant effect of (−)-deprenyl was published by Mann and Gershon (1980).

The realization of the peculiar effect of (−)-deprenyl, first in Parkinson's disease and later in Alzheimer's disease, distracted attention from its antidepressant property, which remained unutilized. Even an especially interesting aspect of this problem fell into oblivion. In a study performed by Birkmayer et al. (1984) on 102 outpatients and 53 inpatients, (−)-deprenyl was given together with (−)-phenylalanine. The latter is a precursor of PEA that, in contrast to PEA, crosses the blood-brain barrier and, as it is metabolized in the brain, increases the concentration of this natural enhancer substance. Nearly 70% of the patients achieved full remission. This outstanding clinical efficiency was equaled only by that of electroconvulsive treatment, but without the latter's side effect of memory loss.

As the drug was primarily used in Parkinson's disease and this illness is very often accompanied by depression, clinicians who treated Parkinson's disease with (−)-deprenyl realized and described the antidepressant effect of the drug in parkinsonism (Youdim 1980; Tom and Cummings 1998; Miyoshi 2001; Zesiewicz et al.,1999).

Quitkin et al. (1984) found (−)-deprenyl effective against atypical depression. This open trial on 17 patients made the finding questionable. But McGrath et al. (1989) in a placebo-controlled trial of (−)-deprenyl proved the efficiency of the drug in atypical depression. In a double blind evaluation Mendlewicz and Youdim (1983) found (−)-deprenyl treatment successful in major depression. Some authors (Lees 1991; Kuhn and Muller 1996; Ritter and Alexander 1997) realized marked antidepressant effect of high doses of (−)-deprenyl (40–60 mg/day). In a study by Bodkin and Amsterdam (2002) the transdermal application of (−)-deprenyl was effective in double-blind, placebo controlled, parallel group examinations in depressed outpatients. Amsterdam (2003) checked the trial and found a good antidepressant effect of (−)-deprenyl. Considering the results of the detailed pharmacological analyses with (−)-deprenyl and with (−)-PPAP, its close structural analogue free of MAO-B inhibitory potency, there is good reason to accept the conclusion that the observed antidepressant efficacy of the drug is unrelated to MAO-B inhibition and that the enhancer effect of the compound is fully responsible for this effect (Knoll 1998).

(−)-BPAP is about 130 times more potent than (−)-deprenyl in rats for antagonizing tetrabenazine-induced depression in the shuttle box. As was discussed previously (Sect. 3.1.3.2), (−)-BPAP, in contrast to (−)-deprenyl, is a highly efficient enhancer of the serotonergic neurons in the mesencephalon. In all the experimental studies performed with (−)-BPAP in comparison to (−)-deprenyl, the tryptamine-derived enhancer substance proved to be substantially more potent. There is good reason to expect that (−)-BPAP will in all likelihood significantly surpass the antidepressant effect of (−)-deprenyl. As is well-known, a considerable percentage of sufferers of major depression cannot be cured with the available antidepressants. Since selective uptake inhibitors, the most effective antidepressants presently used, and enhancer substances stimulate the catecholaminergic and serotonergic neurons in the brain via quite different mechanisms (Miklya and Knoll 2003), their combination with synthetic mesencephalic enhancer substances opens up a new prospect for treating depressed patients in the future.

Parkinson's Disease. The neostriatum is the main input structure of the basal ganglia. We have to consider the physiological role of the nigrostriatal dopaminergic neurons in the continuous activation of the cerebral cortex (see Mink and Thach 1993; Standaert and Young 1996, for review). This is realized via a highly complicated route of connections (see Albin et al. 1989, for review). The neostriatum is the main input structure of the basal ganglia. It gets glutamatergic input from many areas of the cerebral cortex. Cholinergic and peptidergic striatal interneurons are in connection with the nigrostriatal dopaminergic neurons. Dopamine released in the striatum controls the two GABAergic pathways along which the outflow of the striatum proceeds. One is a direct route to the substantia nigra pars compacta and medial globus pallidus. The other is an indirect route. A GABAergic link binds the striatum to the lateral globus pallidus; from here, another GABAergic pathway goes to the subthalamic nucleus, which provides glutamatergic excitatory innervation to the substantia nigra pars compacta and medial globus pallidus; this then continuously inhibits – via a GABAergic projection – the activity of the ventroanterior and ventrolateral nuclei of the thalamus, which provide feedback glutamatergic excitatory impulses to the cerebral cortex.

Thus, the stimulation of the direct pathway at the level of the striatum is increasing the excitatory outflow from the thalamus to the cortex, whereas stimulation of the indirect pathway has the opposite effect. The striatal GABAergic neurons of the direct pathway express primarily the excitatory D_1 dopamine receptors; the striatal neurons of the indirect pathway express primarily the inhibitory D_2 receptors. As a result dopamine release in the striatum increases the inhibitory activity of the direct pathway and diminishes the excitatory activity of the indirect pathway. As a net effect the inhibitory influence of the substantia nigra pars reticulata and medial globus pallidus on the ventroanterior and ventrolateral nuclei of the thalamus is reduced, thus increasing the excitatory effect of these nuclei on the cerebral cortex. All in all, *a more active nigrostriatal dopaminergic system means a more active cerebral cortex and, vice versa,* the physiological age-related decline of the nigrostriatal dopaminergic activity leads to an equivalent reduction in the activity of the cerebral cortex. It is reasonable to conclude that the age-related decline of the nigrostriatal dopaminergic brain mechanism plays a significant role in the decline of performances over time.

Aging of the dopaminergic system in the brain plays an indisputably leading role in the highly significant, substantial decline in male sexual activity and also in the more modest but still significant age-related decline in learning performance. As discussed above (see Sect. 3.5.1.2.1) in a human male study median coital activity was the highest, 2.1 events/week, between the ages of 30 and 34, and decreased progressively with increasing age, sinking to 0.2/week ($P < 0.001$) in the 65- to 69-year-old age group. We found essentially the same trend of changes in male rats in a different series of experiments.

There is a quantitative difference only between the physiological age-related decline of the dopaminergic input and that observed in Parkinson's disease. In the healthy population the calculated loss of striatal dopamine is about 40% at the age of 75, which is about the average lifetime. The loss of dopamine in Parkinson's disease is 70% or thereabout at diagnosis and over 90% at death. The drastic reduction of the dopaminergic output in Parkinson's disease evidently leads to an accordingly drastic reduction of cortical activity and this makes it clear why an enhancer substance, like (−)-deprenyl, improves cognition, attention, memory and reaction times and brings about subjective feelings of increased vitality, euphoria and increased energy in people with Parkinson's disease (see Sect. 3.2.1).

In diagnosing Parkinson's disease, the neurologist selects subjects with the most rapidly aging striatal dopaminergic system (about 0.1% of the population). As symptoms of Parkinson's disease become visible only after the unnoticed loss of a major part (about 70%) of striatal dopamine and further deterioration is irresistible, the disease is, in this sense, incurable. Prevention is the only chance to fight off Parkinson's disease. We need to start slowing the age-related functional decline of mesencephalic enhancer regulation in due time. For this reason it is advisable to begin the prophylactic administration of a synthetic mesencephalic enhancer substance, for example 1 mg (−)-deprenyl/day, as soon as sexual maturity has been reached and the postdevelopmental period of life has just started. It is therefore of particular importance that (−)-deprenyl, to date the only synthetic mesencephalic enhancer substance in clinical use, has proven to be an unusually safe drug.

3.5 Therapeutic Aspects of Synthetic Mesencephalic Enhancer Substances

The DATATOP study in the USA (Parkinson Study Group 1989), the French Selegiline Multicenter Trial (FSMP) (Allain et al. 1991), the Finnish study (Myttyla et al. 1992), and the Norwegian-Danish Study Group (Larsen et al. 1999), all multicenter studies that used (−)-deprenyl as initial treatment in *de novo* patients with Parkinson's disease, have supported the conclusion that (−)-deprenyl slows the progression of early Parkinson's disease and have demonstrated the safety of the long-term administration of the drug. It is commonly assumed that (−)-deprenyl by itself is an exceptionally safe compound.

Due to the inhibition of MAO-B, (−)-deprenyl treatment allows for a 20–50% decrease in the levodopa dose needed in Parkinson's disease. In patients who need levodopa, however, there is always a risk that the administration of (−)-deprenyl will enhance the side effects of levodopa, which can only be avoided by properly decreasing the levodopa dose according to the individual sensitivity of the patient. An example of a multicenter clinical trial in which the improper combination of levodopa with (−)-deprenyl led to confusion and misinterpretation is the one performed by the Parkinson's Disease Research Group of the United Kingdom (PDRG-UK) (Lees 1995).

Quite unexpectedly, this group published an alarming paper claiming that parkinsonian patients treated with levodopa combined with (−)-deprenyl show an increased mortality in comparison with the patients treated with levodopa alone (Lees 1995). This finding was in striking contradistinction to all other studies published in a variety of countries. Birkmayer et al. (1985) even found an increased life expectancy resulting from the addition of (−)-deprenyl to levodopa treatment in Parkinson's disease. The "idiosyncratic prescribing" (Dobbs et al. 1996) of (−)-deprenyl in combination with levodopa in the PDRG-UK study led to the false conclusion of the authors. Comments (Dobbs et al. 1996; Knoll 1996; Olanow et al. 1996) pointed uniformly to the substantial overdosing of levodopa as the cause of the observed deaths with (−)-deprenyl as an adjuvant in this trial.

There are several promising opportunities in the treatment of Parkinson's disease: 1. to slow the progress of the disease and delay the timepoint at which levodopa is needed in *de novo* patients with Parkinson's disease by administering (−)-BPAP, a much more potent and selective enhancer substance than (−)-deprenyl; 2. to administer a carefully adjusted dose of (−)-deprenyl when levodopa is already needed; and 3. to make a safe use of the levodopa-sparing effect of (−)-deprenyl.

Since the diagnosis of Parkinson's disease is the unequivocal proof that the age-related irreversible deterioration of the nigrostriatal dopaminergic neuronal system has already surpassed a critical level in the patient, and the disease is incurable, prevention remains the only chance for the future to fight off Parkinson's disease. The daily administration of a small dose of a synthetic mesencephalic enhancer substance from sexual maturity until death presents itself as a proper method for reaching this aim.

Alzheimer's Disease. Alois Alzheimer described in 1907 the form of dementia that bears his name. He was the first who pointed to a relationship between dementia and the extensive appearance of dense fiber-like tangles and darkly staining senile plaques in the cortical and hippocampal regions. The grave morphological changes lead to grave functional disturbances. For example, the loss of pyramidal neurons and their synapses necessarily leads to cholinergic and glutamatergic hypofunction. As the important role of these transmissions in cognitive and memory functions is well-known, the current symptomatic treatment of Alzheimer's disease is based on the correction of these hypofunctions. But none of these strategies came up to expectations.

Another strategy to treat Alzheimer's disease, the "β-amyloid cascade" theory, is based on neuropathological changes verified postmortem, the excessive appearance of which is claimed to be characteristic of the disease: accumulation of neurofibrillary tangles composed of hyperphosphorylated tau proteins and extracellular senile plaques containing β-amyloid$_{1-40}$ and β-amyloid$_{1-42}$ (see Morishima-Kawashima and Iharra 2002, for review).

As β-amyloid$_{1-42}$ is a neurotoxic agent, the hypothesis that this is a key molecule in the pathology of Alzheimer's disease is now widely accepted (see Selkoe 2001, for review). The theory is controversial since the correlation between the concentrations and distribution of amyloid depositions in the brain and parameters of Alzheimer's disease pathology, such as the degree of dementia, loss of synapses, and loss of neurons, is poor (Neve and Robakis 1998).

As neurotoxicity is thought to be inseparable from oxidative injuries, free radicals, calcium and inflammatory-mediated processes, agents with protective effect on cultured neurons, anti-oxidant compounds, and anti-inflammatory drugs are continuously tested in Alzheimer's disease. For example, vitamin E and selegiline (Sano et al. 1997; Grundman 2000; Thomas 2000; Kitani et al. 2002; Birks and Flicker 2003), Ginkgo biloba extract (Ponto and Schultz 2003), nonsteroidal anti-inflammatory drugs (Etminan 2003), estrogen (Schumacher et al. 2003) have been administered.

Despite the rapid growth of the number of papers that bear evidence of the seemingly frantic success of the "β-amyloid cascade" theory, a survey of the literature furnishes unequivocal evidence that the therapy based on this strategy has not changed the hopelessness of the patients who had already developed Alzheimer's disease. A radically new approach is needed to curb the predicted dramatic increase in the prevalence of Alzheimer's disease.

The first two studies to demonstrate the beneficial effect of (−)-deprenyl in Alzheimer's disease were published in 1987 (Martini et al. 1987; Tariot et al. 1987), and a series of clinical studies with small sample sizes confirmed thereafter the usefulness of this drug in the treatment of the disease (see Knoll 2001, for review). In some of these studies the effect of (−)-deprenyl was compared with other drugs. Campi et al. (1990) found (−)-deprenyl to be more effective than acetyl-L-carnitine. According to Falsaperla et al. (1990), (−)-deprenyl is more effective than oxiracetam (a piracetam-like nootropic drug) in improving higher cognitive functions and reducing impairment of daily living. In a study by Monteverde et al. (1990) (−)-deprenyl proved to be more effective than phosphatidylserine.

The rationale and design of the first multicenter study of (−)-deprenyl in the treatment of Alzheimer's disease using novel clinical outcomes was published by Sano et al. in 1996 and the results of this study were published 1 year later (Sano et al. 1997). The primary outcome involved the time that elapses until the occurrence of any of the following: death, institutionalization, loss of the ability to perform basic activities of daily living, or severe dementia. There were significant delays in the time taken for such primary outcomes to occur in patients treated with (−)-deprenyl. The authors concluded that in patients with moderately severe impairment from Alzheimer's disease, treatment with (−)-deprenyl slows the progression of the disease.

*

The operation of mesencephalic enhancer regulation provides a true perspective on the development of the two main neurodegenerative diseases (Parkinson's, Alzheimer's). As has already been discussed above in connection with Parkinson's disease, a more active brain engine means a more active cerebral cortex. Accordingly, the mesencephalic brain engine keeps the telencephalic neurons continuously active in the developmental phase of life and in the early phase of postdevelopmental longevity. These neurons fight successfully against their innate self-produced harmful metabolites. Due to the irresistible, slow decay of mesencephalic enhancer regulation as a function of age, the beneficial influence of the brain engine on the telencephalic neurons progressively declines. As a result of this process, with the passing of time even morphologically traceable age-related neurodegenerative changes appear in the highly sensitive, most sophisticated telencephalic neurons.

Alzheimer's disease – an irreversible loss of neurons primarily in the cortex and hippocampus leading to progressive impairment in memory, judgement, decision making, and so on – is the worst outward form of brain aging. An analysis of the prevalence of Alzheimer's disease as a function of age makes it clear that this is just a grave form of the natural aging of the human brain.

The mean age at the onset of Alzheimer's disease is approximately 80 years, and the manifestation of the illness before the age of 60–65 years is very rare. In the age cohort 65–69, Alzheimer's disease has a prevalence of only 1%. This increases to about 20% in the 85- to 89-year-old group and the risk of precipitating the disease can reach the 50% level among persons 95 years of age and over (Campion et al. 1999; Hy et al. 2000; Helmer et al. 2001; Nussbaum and Ellis 2003). The prevalence of Parkinson's disease over the age of 80 is only 1–3% (Tanner et al. 1996).

In the population over 65, there are substantial sex (68% female, 32% male) and geographical (2.1% Japan, 5.2% Europe and 10% USA) differences in the incidence of Alzheimer's disease (see Lockhart and Lestage 2003, for review). The disease now affects about 15 million persons worldwide, and a sharp increase in the afflicted population is expected in the future as it is estimated that the number of individuals over 65 will increase to 1.1 billion by 2050. It is therefore a pressing and no longer postponable necessity to find a safe and efficient prophylactic therapy to significantly decrease the prevalence of Alzheimer's disease as soon as possible.

There are special genetic risk factors for Alzheimer's disease – such as, for instance, the $\varepsilon 4$ allele of the apolipoprotein E (APOE) gene, isotonic variation in CYP46, CYP46*C – that significantly increase the risk of Alzheimer's disease development (Bookheimer et al. 2000; Borroni et al. 2004). Accordingly, it seems reasonable to assume that the majority of those who precipitate the disease are carriers of risk factors.

The only reasonable hope to fight off Alzheimer's disease in the future is to keep the cortical and hippocampal neurons at a higher activity level as long as possible by the prophylactic administration of a synthetic mesencephalic enhancer substance. It is remarkable in this regard that (−)-BPAP protected cultured rat hippocampal neurons from the deleterious effect of β-amyloid$_{25-35}$ fragments in as low as 10^{-15} M concentration (Knoll et al. 1999).

4 Approaching Old Problems From A New Angle

4.1
A New Interpretation of the Substantial Individual Differences in Behavioral Performances

It is a common experience that – independent of method and species used for studying learning-based behavioral modification – the observer runs into substantial individual differences in performance, for reasons unknown. Conceiving learning as a cortical enhancer regulation dependent function, as discussed in Sect. 3.4.2, and recognizing that the enhancer substance exerts its effect in terms of a peculiar bell-shaped dose/response curve (see Fig. 3.11) offers a reasonable interpretation for this remarkable phenomenon.

As shown in Fig. 3.11, the most effective dose of (−)-BPAP, 0.0005 mg/kg, increased the release of norepinephrine from 4.7 ± 0.10 nM/g (control) to 15.4 ± 0.55 nM/g ($P < 0.001$), but a 100-times higher dose of (−)-BPAP (0.05 mg/kg) did not change it (4.3 ± 0.25 nM/g) (Knoll et al., 2002b). Since an *optimum* concentration of the enhancer substance was needed for the *optimum* performance, *I postulate that the substantial individual differences found in behavioral performances are due to the peculiar dose-dependency of the endogenous mesencephalic and cortical enhancer substances.*

This approach grants us a new perspective on the results of our two longitudinal studies on rats (Knoll 1988; Knoll et al. 1989, 1994). As an example, let us analyze our second longitudinal study (Knoll et al. 1994) from this perspective. We selected the sexually highest- and lowest-performing individuals from a population of 1,600 rats.

We started working with a random population of 28-week-old male rats and tested their sexual performance once a week. Rats representing the two extremes in performance were selected for the study: ones that did not display a single intromission during the four consecutive weekly-mating tests used for selection, and ones which showed full-scale sexual activity (mounting, intromission, ejaculation) in each of the four tests. Out of 1,600 sexually inexperienced 28-week-old Wistar-Logan male rats that met a receptive female once a week for 4 consecutive weeks, 94 did not display a single intromission during the selection period and 99 displayed at least one ejaculation in each of the four tests. The former were taken to be the sexually lowest performing (LP) rats, and the latter the highest performing (HP) ones.

After selection, rats were treated subcutaneously with either 1 ml/kg 0.9% NaCl or with 0.25 mg/kg (−)-deprenyl, dissolved in 0.9% NaCl given in the same volume, three times a week, to the end of their life. Out of the 94 LP animals, 46 were treated with salt. Out of the 99 HP animals, 49 were treated with salt. The performance of these salt-treated LP and HP rats during a testing period of 108 weeks is shown in Fig. 4.1. The figure shows: (a) the average of the total number of ejaculations displayed in the weekly mating tests; (b) the average of the total number of CARs produced in the learning tests (The rats were trained once every 3 months for a period of 5 days, with 20 trials a day in the shuttle box); and (c) the average lifetime of the rats.

Figure 4.1 demonstrates the highly significant difference in sexual and learning performances and in life span between LP and HP rats.

Considering the unique dose-related effect of an enhancer substance, as shown in Fig. 3.11, we assume that out of the 1,600 rats the 99 HP rats produced their endogenous enhancer substances at the peak of the bell-shaped concentration/effect curve, while the 94 LP rats produced them at the least active part of the curve; and the production of the overwhelming majority of the population (1,407 rats) falls between the two extremes.

An analysis of the ability of rats to acquire the glass-cylinder-seeking drive is another example that convincingly illustrates the great individual differences

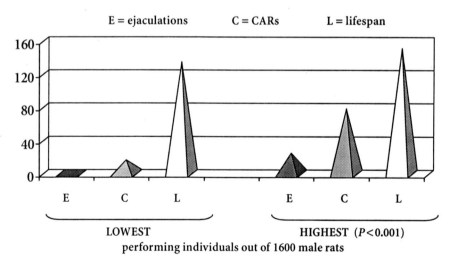

Fig. 4.1. Illustration of the highly significant differences in sexual and learning performances and in life span between two groups of rats selected from 1,600 28-week-old Wistar-Logan males, as the sexually lowest performing (LP) and highest performing (HP) individuals. See text for details

4.1 Individual Differences in Behavioral Performances

in the behavioral performances of rats. As discussed earlier (see Sect. 1.3), despite the same training procedure, only 2 rats out of 100 acquired a lifelong operating glass-cylinder-seeking drive. Presumably only these two glass-cylinder-seeking rats were lucky enough always to mobilize, in the optimum concentration, their specific endogenous enhancer substances in the cortical neurons responsible for the operation of the glass-cylinder-seeking drive.

*

Acquired drives are the primary determinants of human activity. Whatever humans have discovered, invented, developed, renewed, or modified, has stemmed from previously attained knowledge, from chains of ICRs previously irreversibly fixed in their cortex through innate or acquired drives. Very hard work is needed to accumulate proper knowledge in the brain. Even creative work in science or art can originate only from that which is already known. The learned is digested, rethought, varied until the new problem to be solved is shaped in the brain of the creative mind such that a ripe new variant ultimately results.

The case of Mozart is a convincing example that this is true even if a genius creates something astounding as a child. From his birth, Wolfgang Mozart participated passively, but continuously, in the process of teaching music. He was, namely, always present when his father taught his elder sister. He himself was then taught with strict discipline from his third year of age onward. Leopold Mozart, who himself was a better-than-average musician, set himself as the main objective of his life to educate his son for a one of a kind career as musician. It is hard to overestimate the role of these special circumstances during his early years in Wolfgang Mozart's exceptional accomplishment. Without the unusually hard exercise regimen imposed upon him by his father, he would not have been able to fix all those millions of chains of ICRs that formed the foundations of his brilliant creations. Moreover, he traveled with his father as a young boy, became acquainted with the creations of the best musicians of his age, including the music of Lully, Händel and Haydn, and even knew Haydn personally.

He was considered to be an infant prodigy who immediately learned to create his own variations, renewals; his own compositions. Since creation can originate only from the already known, Mozart composed sonatas, operas, symphonies, serenades, masses, and so on with the greatest of ease. With similar ease he learned to play any kind of musical instrument, and was only about 7 years old by the time he had mastered the most complicated instrument, the organ. As a wonder boy he performed in the palaces of the élite everywhere with great success, and, most importantly, he was also adored by the aristocrats in Vienna, and even by Queen Maria Theresa and her son, Joseph. But as he grew elder, and in order to exploit his unique capabilities, he tried to be an independent musician, but was refused everywhere. The social circumstances in his time and native country were not suitable for such ambitions (about twenty years later Beethoven successfully reached this aim). Mozart, coming into conflict with himself and losing his lust for life, died prematurely.

*

Cortical enhancer regulation, the proposed key mechanism of learning (see Sect. 3.4.2), is the same in every brain. Regarding any performance, there is only a quantitative difference between the worst and the best performing individual. The worst performing individual synthesizes when needed, its specific enhancer substance in an ineffective concentration, while the best performing one is producing it in its most active concentration. Thus, according to this approach, talent signifies an innate endowment such that cortical neurons responsible for a given performance are always capable of mobilizing, when needed, their specific enhancer substance at optimal concentration. Because of the infinite individual differences in the natural endowments of the billions of cortical neurons, it is clear that everybody tends to acquire the best-fitting drives. These drives are those which are related to cortical neurons with the capacity to produce when needed the critical enhancer substance in optimal concentration. However, as the life of a human being depends primarily on the living conditions offered by the micro-society to which the person belongs, it is always unpredictable how successful an individual can be in the acquisition of work-related drive(s) to his/her best needs. Serious psychic problems are obviously rooted in uncorrectable living conditions that hinder the realization of ambitions. Accordingly, optimally performing humans are satisfied, well-balanced and happy, frustrated ones are lost in pseudo-activities and search for "ersatz": smoking, drinking, drugs, gambling, overeating, pursuing excessive sexual pleasures, and so on.

4.2
A New Interpretation of Forgetting, Remembering, and Boredom

According to our concept learning means the development and irreversible manifestation of a specific cooperation between cortical neurons that is based on the operation of cortical enhancer regulation. The cooperating neurons work as an integrated whole. The activation of this entity means that in response to proper stimulation enhancer regulation in the neurons starts working at a higher activity level and the cooperating neurons reach a level of excitability needed for the precise execution of the learned performance. When for example a long chain of ICRs is elicited, the cooperating millions of cortical neurons get explosively excited and the row of engrams is ecphorized in the sequence in which they were stored during the learning process.

Whenever we study how, in the long run, an already irreversibly fixed information, a highly complicated chain of ICRs can later be ecphorized, we face the problem that

4.2 A New Interpretation of Forgetting, Remembering, and Boredom

1. Unexpectedly, member(s) of the complicated chain cannot be ecphorized when needed (forgetting).
2. In the course of further trials the temporarily forgotten information can be recalled again (remembering).
3. Sooner or later, after tedious repetitions of the complicated chain of ICRs, a boredom-like modification of behavior regularly appears.

To penetrate into the essence of forgetting and remembering in the light of the enhancer regulation concept, let us think about the changes in the brain when we recall four lines from a poem that we learned decades ago and know by heart.

A poem is a set of words in a precisely fixed order. To fix a poem in the cortical neurons and recall it faultlessly as an entity at any later date means the activation of a chain of ICRs, i.e., millions of cortical neurons are activated in the same sequence as they were activated at the time when the chain was fixed in the brain by training.

Let us recall the following four lines of a poem by Auden:

> She was my North, my South, my East and West
> 1   2   3   4   5   6   7   8   9   10
> My working week and my Sunday rest
> 11   12   13   14 15   16   17
> My noon, my midnight, my talk, my song
> 18   19   20   21   22 23   24 25
> I thought that love would last for ever, I was wrong.
> 26   27   28   29   30   31 32   33   34 35   36

The stanza of the poem quoted here consists of 36 words. Each word by itself is a chain of ICRs consisting of letters that were fixed in an exact order forever when we learned the language. The cited text, a highly complicated chain of ICRs, consisting of 36 words, was fixed in exact order when we learned the poem.

For the sake of perspicuity, let us think about the changes in the brain when we recall, for example, word 21 – *midnight* – from the poem.

The word

> m   i   d   n   i   g   h   t
> 1   2   3   4   5   6   7   8

consist of eight letters, which were fixed as a chain of ICRs when we learned the language. Whenever we recall the word – the eight members (letters) of the chain, each fixed in a special group of cortical neurons – we activate, in the proper sequence, groups of millions of cortical neurons that once in the past had learned to work together. Due to the explosive activation process, the word, as an integral whole, is consciously perceived.

A word is an elementary unit serving interpersonal communication. It is a relatively simple chain of ICRs. Being an important tool, it is usually faultlessly ecphorized whenever needed. But, because of the natural, unpredictable, substantial undulation of cortical enhancer regulation, this does not mean that whenever we recall the word each unit of the chain, each letter, reaches the level of excitability needed for the conscious recalling of the word as an entity. But this is not necessary! It is enough to recall the first, the last and some of the letters in between to ecphorize the word as an integral whole. This can be easily checked. When, for instance, we read the word *mednight* in a text, we will read it correctly, usually without noticing the misspelling.

We assume that the phenomenon of forgetting and remembering is attributed to the continuously changing excitability of the cortical neurons, due to the physiological undulation in the synthesis of the specific enhancer substance. The operation of this mechanism always makes it unpredictable which group of neurons – in the course of the explosive activation of a chain of ICRs – will be temporarily unable to reach the level of excitability needed for conscious perception.

Considering this mechanism, the common experience is understandable that we may finally bring up the forgotten word by recalling the whole line of the verse several times in succession. Due to the undulation of cortical enhancer regulation, the group of cortical neurons responsible for the missing word, which initially failed to produce the critical amount of the specific enhancer substance needed for the conscious perception of the engram, ultimately reaches by chance the required state of excitability, and we *remember* the word. Yet the infallible method of immediately calling back the word is to read the line again. In this case we activate the cortical neurons in a nonspecific manner via the catecholaminergic system in the brain stem.

Anyone who has learned poems by heart and is used to recalling them for years and years, though only occasionally, can easily detect a continuously operating mechanism in the brain which counters the consequences of forgetting. When after years we check the original wording of the poem against the one in our memory, we will regularly find alteration(s) that did not change the poem's basic meaning. Furthermore, we will find it difficult to eliminate the fixed substitute(s) and return to the original wording.

It seems logical that when we recalled the poem from memory some of the original words were *forgotten* and were automatically substituted for by synonyms, adjusting correctly for the forgotten ones, and the new variants were fixed in an inextinguishable manner.

I for example had once learned the stanza from Auden quoted above and had originally fixed it faultlessly in my memory. I recalled the stanza occasionally without reinforcement for years, but read the original text again only

4.2 A New Interpretation of Forgetting, Remembering, and Boredom

several years later. I realized that with the passing of time the wording was automatically modified in my brain. Instead of the originally learned words 12 and 13 – *working week* – I recalled a well-fitting *working days* variant.

Considering the described mechanism of forgetting, we can reasonably explain the modification of the text as follows. By recalling the four lines of the Auden poem from memory, the members of the chain, consisting of 36 words, activated each other in the fixed sequence. In contrast to others, the group of neurons representing word 13 once remained below the excitability needed for conscious perception. Though word 13 properly stimulated word 14, and so on, I temporarily forgot word 13 and, for me as a nonnative speaker, the never used *working week* was automatically replaced by the trivial *working days*. It is obvious that this highly sophisticated mechanism plays a crucially important role in countering the failures caused by the unpredictable undulation of enhancer regulation in the cortical neurons.

The fact that interchangeable words with equivalent meaning are stored in the brain is of great practical importance in countering the unavoidable age-related decline of cortical enhancer regulation. Let us consider that when a child learns a word, a unit of a language, a symbol having a simple referent or representing an indivisible concept, action, or feeling, the attraction between the proper cortical neurons is fixed for a lifetime, as this is the main condition for interpersonal communication. A word, consisting of letters of the alphabet, is a chain of ICRs. Each letter is by itself a chain of ICRs. The sequence of the letters determines the word which is recalled as an entity. It is obvious that when this chain of ICRs was fixed, the neurons which got to know each other and thereafter worked in concert whenever the word was used are widespread in the cortex. But we also have synonyms, other words with essentially the same meaning fixed in a quite different group of cortical neurons. Thus, we have a couple of interchangeable words. Those who speak another language also have another group of symbols with the same meaning. Each of these words is fixed in separate groups of cortical neurons. However, irrespective of where they are located, as all these symbols have the same meaning, they can be used as interchangeable tools, because they play the same role in interpersonal communication.

It is well known that people who leave their native country and learn a new language in fact mix the words of their mother tongue with the ones of the new language that they are learning. On the other hand, after not using their mother tongue for a long time, they will mix in the words of their newly acquired language when communicating in their native language.

Evidence that words with the same meaning are fixed in the brain in different cortical neurons is beautifully provided by patients who speak more than one language, suffer a mild stroke resulting in the transient loss of their ability to speak and are compelled to communicate in writing. For example, a cultured, highly qualified Hungarian expert who fluently spoke English, temporarily lost his ability to speak and communicated in writing. He wrote his messages flowingly in his mother tongue, in Hungarian, but used a few English words. This was clear proof that he had, at least for the time being, lost the ability to use the whole vocabulary of his mother language. Because of the transient inability to ecphorize some words of his mother tongue, he replaced them immediately with the equivalent English symbols that were obviously stored in uninjured cortical neurons. After a couple of hours the normal performance of his cortex came back and the patient gave account of his conscious experience that he was unable to ecphorize some Hungarian words and was compelled to use the English equivalents to convey his message.

As cortical enhancer regulation is necessarily also subject to an age-related decline, it is obvious that during the act of recalling a chain of ICRs, the number of members in the chain that are temporarily unable to reach the level of excitability necessary for conscious perception gets higher and higher with the passing of time. Thus, the protective mechanism that the forgotten word can be replaced with an equivalent variant is of peculiar importance for the aging brain. The higher the brain's potential to provide substitutes for temporarily forgotten words, the smoother the person's adaptation to the unavoidable age-related decline in cortical performance.

*

Getting bored is a common human experience. The two diametrically opposite reasons that lead to boredom are inactivity and the tedious repetition of any form of previously well-trained activity.

In the introduction to his best-known work, *Les fleurs du mal*, Baudelaire described boredom as the most dreadful enemy of the human soul, and thousands of others have written vividly about this phenomenon. We can easily find in famous novels exciting examples of romantic heroes whose life stories clearly illustrate the strange significance of boredom in a human's life. Biographies of creative artists allow us to get insight into the sometimes fatal consequences of the operation of this physiological phenomenon on highly sensitive, creative, human brains.

In our studies with glass-cylinder-seeking rats, we saw that once the animals manifested the acquired drive, they searched for the glass cylinder repeatedly and for long periods of time without any signs of trouble. As time passed, however, tedious repetitions of glass-cylinder search efforts in an unchanged environment led to a peculiar behavioral modification. After eliciting 10–20 glass-cylinder searches with 30-s or 1-min intervals daily, in an unchanged environment and for a longer period, the rat, which had previously performed well, started to linger as a rule for a longer time at one of the positions where a higher impetus was needed to overcome resistance. The appearance of this behavioral pattern bore striking resemblance to weariness.

4.2 A New Interpretation of Forgetting, Remembering, and Boredom

We studied this phenomenon in a setup shown in Fig. 4.2. The rat is placed on a plate, 60 cm in diameter (SPOT 1) from which a bridge (A) 1-m long and 30-cm wide leads to a 2-m long, 30-cm wide and 60-cm high wooden bench (B). At the end of this bench the rat arrives at SPOT 2, at the bottom of a metal rod (C) 0.5 cm in diameter and 90-cm high. The rat is compelled to climb the rod to reach a wooden plate (D) 1.5-m long and 20-cm wide. This wooden plate is connected to a metal plate (SPOT 3) on which a glass cylinder – 30-cm high, open on bottom and top with diameters of 16 cm and 12 cm – is placed. The glass cylinder has a side opening through which a rat up to 350–400 g body weight can conveniently manage to get inside and jump onto the top of it.

The rats were trained to acquire the drive to find the glass cylinder in an unknown environment and jump on the top of it. The phenomenon strikingly reminiscent to boredom appeared in rats that were compelled, after the acquisition of the glass-cylinder-seeking drive, to search for the glass cylinder at least 20 times daily in an *unchanged* environment for a longer period of time. As a consequence of this form of training, the characteristic change in behavior was already observable in some of the well-performing rats within

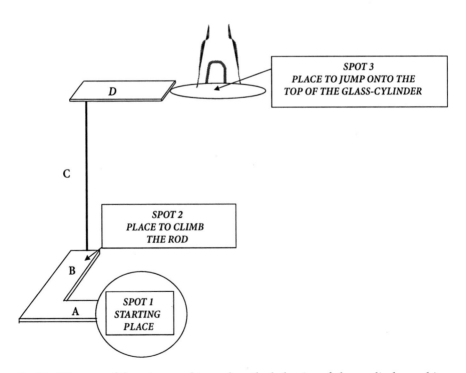

Fig. 4.2. Diagram of the setup used to analyze the behavior of glass-cylinder-seeking rats (see details in the text)

3–4 weeks, though with the others months passed until the phenomenon appeared. Protocols 1, 2, and 3 illustrate the characteristics of the special change in the behavior of glass-cylinder-seeking rats that is reminiscent of what we term *boring* in humans.

In the Protocols the time course of a successful performance of the rat from start until the animal jumped onto the top of the glass cylinder is divided into five consecutive stages:

1. Interval between activating the shrill bell (conditioned stimulus) and the rats moving away from SPOT 1

2. Time taken to reach SPOT 2

3. Time taken to climb the rod and reach the wooden plate at the upper end of the rod

4. Time taken to run to SPOT 3

5. Interval between arriving at SPOT 3 and jumping onto the top of the glass cylinder

The rat with the performance detailed in Protocol 1 behaved as usual on March 4 and 5. The duration of the first reaction, as always, was a longer one on both March 4 (12′01″) and March 5 (10′27″). This is the characteristic "warming up" behavior of the glass-cylinder-seeking rats (see Knoll 1969, for review). A longer time is obviously needed until the state of enhanced excitability is reached in the proper group of cortical neurons which operate as the "active focus," the cortical representation of the drive.

The typical behavior resembling being bored appeared unexpectedly on March 6. The rat lingered a long time before it started to work in the setup. To illustrate this change: on March 4 and 5 the duration of the 10th performance was 57″ and 37″, respectively, while it lasted 9′21″ on March 6. After the 10th response we changed the environment. The plate from which the rat used to start and where it lingered a long time was eliminated. The animal was placed on the bridge (A) that leads through the wooden bench (B) to SPOT 2.

This simple change had a dramatic effect. The rat ran through the whole setup, reached the glass cylinder and jumped on the top of it within 38″! The rat never performed before with such speed, and it worked thereafter with unflagging zeal during the whole experiment. The 18th and 19th performance lasted altogether 17 and 15 s, respectively, which is an almost unbelievable speed for running through the entire cycle (bench, climb the rod and reach the glass cylinder). This was the best performance we had ever seen in this setup from among dozens of excellently performing glass-cylinder-seeking rats. Thus, the change of the already boring environment excited the rat immediately with an intensity never observed before.

4.2 A New Interpretation of Forgetting, Remembering, and Boredom

Protocol 1. Illustration of the typical behavior of a rat which proved to be highly successful in the acquisition of the glass-cylinder-seeking drive. The rat fixed the new drive after an unusually short, daily training period of only 3 weeks. It soon displayed, however, the typical behavioral pattern strikingly resembling boredom. The training period of the rat started on February 10. The protocol demonstrates the performance of the rat on March 4 and 5. These were the last two days that preceded the first appearance of weariness on March 6. For sake of a better overview, the time course of the whole glass-cylinder-seeking performance is shown in the five consecutive stages described in the text

Date of experiment	No. of the consecutive glass-cylinder searches elicited with 30" intervals	Duration of the whole performance	Duration of stage				
			1	2	3	4	5
4 March	1	12'01"	3'10"	1'50"	5'18"	1'40"	3"
	2	4'29"	25"	10"	3'42"	10"	2"
	3	2'14"	20"	5"	1'42"	5"	2"
	4	1'01"	13"	3"	38"	5"	2"
	5	1'16"	11"	8"	47"	8"	2"
	6	1'17"	5"	8"	59"	4"	1"
	7	1'38"	5"	17"	1'11"	4"	1"
	8	2'35"	5"	8"	2'10"	8"	4"
	9	1'16"	5"	22"	37"	11"	1"
	10	57"	5"	13"	29"	9"	1"
5 March	1	10'27"	4'45"	1'17"	4'18"	5"	2"
	2	1'16"	34"	8"	27"	5"	2"
	3	40"	8"	5"	21"	4"	2"
	4	54"	32"	5"	11"	4"	2"
	5	1'14"	18"	7"	39"	8"	2"
	6	44"	21"	6"	10"	5"	2"
	7	1'59"	1'17"	5"	22"	13"	2"
	8	55"	8"	22"	8"	12"	2"
	9	2'45"	1'42"	17"	39"	5"	2"
	10	37"	13"	5"	12"	5"	2"
6 March	1	10'03"	9'41"	5"	10"	5"	2"
	2	4'58"	4'38"	5"	8"	5"	2"
	3	3'54"	3'08"	17"	22"	5"	2"
	4	7'05"	6'18"	8"	29"	8"	2"
	5	6'20"	5'42"	9"	13"	11"	5"
	6	4'39"	4'08"	8"	12"	5"	6"
	7	6'52"	6'07"	7"	23"	4"	11"
	8	5'02"	4'14"	8"	24"	5"	11"
	9	7'50"	7'18"	5"	18"	4"	5"
	10	9'21"	8'56"	5"	11"	7"	2"
	11	38"	12"	7"	12"	5"	2"
	12	20"	4"	3"	6"	5"	2"
	13	20"	5"	3"	5"	4"	3"

Date of experiment	No. of the consecutive glass - cylinder searches elicited with 30″ intervals	Duration of the whole performance	Duration of stage				
			1	2	3	4	5

We changed the environment after the 10th glass-cylinder-seeking performance. The plate from which the rat started to work and where the animal lingered for a long time was eliminated and the rat was placed directly on the bridge (A)

Date of experiment	No.	Duration	1	2	3	4	5
6 March	14	47″	17″	11″	8″	7″	4″
	15	23″	4″	5″	8″	4″	2″
	16	27″	6″	5″	11″	3″	2″
	17	26″	3″	8″	6″	7″	2″
	18	17″	3″	4″	5″	3″	2″
	19	15″	2″	3″	5″	3″	2″
	20	28″	14″	3″	6″	3″	2″

Experiment performed in the boring environment

7 March	1	11′47″	11′12″	10″	18″	5″	2″
	2	5′11″	4′42″	8″	12″	7″	2″
	3	4′17″	3′54″	6″	7″	6″	4″
	4	4′32″	4′12″	5″	8″	5″	2″
	5	7′57″	7′16″	13″	21″	5″	2″
	6	7′22″	6′36″	11″	28″	5″	2″
	7	4′01″	3′18″	18″	18″	5″	2″
	8	5′20″	4′17″	6″	39″	13″	5″
	9	3′30″	2′18″	17″	43″	8″	4″
	10	4′12″	3′48″	8″	9″	5″	2″

Changing of the environment as on the previous day.

	11	57″	31″	11″	8″	5″	2″
	12	12″	2″	2″	3″	3″	2″
	13	27″	2″	2″	17″	4″	2″
	14	18″	2″	5″	6″	3″	2″
	15	20″	2″	8″	5″	3″	2″
	16	20″	2″	7″	6″	3″	2″
	17	51″	2″	11″	31″	5″	2″
	18	21″	7″	3″	6″	3″	2″
	19	16″	2″	3″	6″	3″	2″
	20	48″	22″	5″	14″	5″	2″

We repeated the experiment on the next day (March 7). The rat started to work in the old, already boring setup, and lingered at SPOT 1 again for a long time. After the 10th performance we changed the environment again as we had on March 6. Protocol 1 demonstrates that the course of the experiment was essentially the same as the day before.

4.2 A New Interpretation of Forgetting, Remembering, and Boredom

Protocol 2. Illustration of the typical behavior of a rat that acquired the glass-cylinder-seeking drive within 40 days, being trained daily. The rat, whose training started on January 21, displayed for the first time the typical signs of boredom on June 3. The protocol shows, as an example, the normal performance of the rat on April 1. The appearance of boredom and its elimination by changing the environment is shown on June 3, and a repetition of this experiment more than one month later, on July 8

Date of experiment	No. of the consecutive glass-cylinder searches elicited with 30″ intervals	Duration of the whole performance	Duration of stage				
			1	2	3	4	5
1 April	1	5′22″	4′12″	40″	18″	7″	5″
	2	1′12″	44″	15″	7″	5″	1″
	3	53″	28″	7″	12″	4″	2″
	4	46″	7″	8″	22″	7″	2″
	5	32″	12″	7″	6″	5″	2″
	6	22″	6″	5″	5″	4″	2″
	7	20″	3″	5″	7″	4″	1″
	8	22″	4″	4″	6″	7″	1″
	9	30″	4″	7″	6″	11″	2″
	10	35″	6″	9″	13″	4″	3″
3 June	1	10′06″	3′18″	7″	6′35″	4″	2″
	2	9′14″	47″	11″	8′09″	5″	2″
	3	10′26″	1′02″	9″	9′09″	4″	2″
	4	7′45″	17″	10″	7′11″	5″	2″
	5	11′49″	6″	8″	11′22″	7″	6″
	6	9′39	15″	9″	9′08″	5″	2″
	7	8′17″	21″	6″	7′42″	5″	3″
	8	23′01″	3″	4″	22′46″	6″	2″
	9	6′57″	5″	13″	6′33″	4″	2″
	10	28′06″	32″	6″	27′21″	5″	2″

We changed the environment. We placed colored paper strips at the bottom of the rod, where the rat lingered before it started to climb the rod.

	11	47″	8″	5″	27″	5″	2″
	12	33″	17″	3″	6″	5″	2″
	13	20″	5″	3″	5″	5″	2″
	14	27″	8″	7″	6″	4″	2″
	15	30″	11″	5″	8″	4″	2″
	16	29″	7″	3″	13″	4″	2″
	17	23″	5″	5″	7″	4″	2″
	18	25″	3″	8″	7″	5″	2″
	19	20″	3″	5″	5″	5″	2″
	20	32″	10″	5″	9″	6″	2″

Protocol 2. (continued)

Date of experiment	No. of the consecutive glass - cylinder searches elicited with 30" intervals	Duration of the whole performance	Duration of stage				
			1	2	3	4	5
Experiment performed in the boring environment							
8 July	1	12'36"	4"	7"	12'18"	5"	2"
	2	6'37"	5"	11"	6'06"	13"	2"
	3	9'57"	31"	3"	9'16"	5"	2"
	4	8'29"	12"	2"	8'08"	5"	2"
	5	5'11"	8"	3"	4'47"	11"	2"
	6	9'38"	27"	2"	9'02"	5"	2"
	7	8'49"	31"	5"	8'06"	5"	2"
	8	7'32"	3"	11"	7'11"	5"	2"
	9	5'54"	3"	2"	5'42"	5"	2"
	10	6'11"	5"	2"	5'35"	7"	2"
We changed the environment as on June 3.							
	11	34"	8"	2"	15"	7"	2"
	12	22"	7"	3"	6"	4"	2"
	13	25"	11"	3"	5"	4"	2"
	14	34"	20"	3"	4"	5"	2"
	15	16"	5"	2"	4"	4"	1"

We studied the appearance of this phenomenon on many well-performing glass-cylinder-seeking rats. We found that the rats lingered either at the start (Protocol 1 gives an example), hesitated to climb the rod (Protocol 2 provides an example), or were unwilling to jump onto the top of the glass cylinder (Protocol 3 furnishes an example).

To find a reasonable explanation for the peculiar phenomenon demonstrated in these protocols the following has to be considered:

1. The rats lingered at positions where a much higher impetus was needed to overcome the resistance and continue towards the glass cylinder. It is obvious that higher impetus is needed to start operating (Stage 1), to climb the rod (Stage 3), or to jump onto the top of the glass cylinder (Stage 5) than to run from Spot 1 to Spot 2 (Stage 2), or to run from the upper top of the rod to Spot 3 (Stage 4). This means that the excitement of the proper cortical neurons, and as a reflection of it – the state of enhanced activity of the brain engine – fluctuates and runs high in a glass-cylinder-seeking rat.
2. We never observed a phenomenon reminiscent of boredom in connection with innate drives, where the inexhaustible mesencephalic neurons keep the cortical neurons active. It therefore seems that tedious repetitions of

4.2 A New Interpretation of Forgetting, Remembering, and Boredom

Protocol 3. Illustration of the typical behavior of a rat which acquired the glass-cylinder-seeking-drive within 58 days, being trained daily. The rat, whose training started on February 16, displayed the typical signs of boredom on July 18 for the first time. The protocol shows as examples: the normal performance of the rat on April 25, the appearance of boredom, its elimination by changing the environment on July 18, and its return to the boring environment on September 17

Date of experiment	No. of the consecutive glass - cylinder searches elicited with 30″ intervals	Duration of the whole performance	Duration of stage				
			1	2	3	4	5
25 April	1	5′14″	4′43″	3″	8″	7″	13″
	2	1′37″	1′04″	4″	8″	3″	18″
	3	1′00″	37″	4″	11″	3″	5″
	4	44″	18″	5″	6″	4″	11″
	5	48″	17″	3″	5″	17″	6″
	6	32″	17″	3″	3″	4″	5″
	7	1′04″	32″	8″	3″	4″	17″
	8	1′44″	1′06″	11″	3″	5″	19″
	9	2′21″	2′02″	3″	5″	7″	4″
	10	1′01″	41″	3″	5″	5″	7″
18 July	1	8′23″	2′02″	3″	4″	3″	6′11″
	2	5′18″	5″	8″	4″	3″	4′58″
	3	7′46″	13″	10″	8″	3″	7′12″
	4	3′45″	8″	5″	11″	3″	3′18′
	5	2′19″	8″	3″	6″	3″	1′59″
	6	6′30″	7″	3″	6″	3″	6′11″
	7	10′32″	5″	3″	13″	3″	10′08″
	8	27′42″	3″	3″	6″	5″	27′25″
	9	8′53″	2″	2″	4″	3″	8′42″
	10	13′15″	2″	3″	5″	3″	13′02″

We changed the environment. We placed a piece of blue cardboard under the glass-cylinder, the place where the rat lingered before it jumped onto the top of the glass-cylinder.

	11	35″	2″	3″	5″	4″	21″
	12	24″	3″	3″	6″	3″	9″
	13	56″	21″	9″	8″	5″	13″
	14	41″	19″	6″	5″	3″	8″
	15	27″	7″	4″	4″	5″	7″

Experiment performed in the boring environment

17 Sept.	1	12′06″	44″	8″	5″	3″	11′06″
	2	9′05″	13″	10″	31″	4″	8′07″
	3	12′32″	12″	3″	8″	7″	12′02″
	4	6′31″	12″	3″	4″	7″	6′05″
	5	3′13″	20″	3″	4″	4″	2′42″
	6	3′08″	8″	5″	3″	4″	2′48″
	7	26′29″	10″	7″	3″	3″	26′06″
	8	21′24″	9″	17″	3″	3″	21′02″
	9	3′25″	7″	6″	3″	3″	3′06″
	10	5′22′	6″	5″	5″	4″	5′02″

glass-cylinder searches with 30 s intervals in an unchanged environment sooner or later lead to a decline of the specific stimulation-induced enhanced excitability in the sensitive group of cortical neurons responsible for the glass-cylinder-seeking behavior. Thus, the situation seems to be essentially similar to the one we analyzed in connection with forgetting a word when we recall a line of a verse that we know by heart. In this case too the nonspecific activation of the cortical neurons, via stimulation of the catecholaminergic system in the brain stem, acts dramatically.

It is perplexing to realize that, in essence, the phenomena we found in rats is retraceable in humans. Let us take, as an example, one of the most sophisticated acquired drives in humans: the alliance by love, based essentially on the innate sexual drive. Human love as an acquired drive is an extremely complicated, unfathomable set of sentimental ties that are continuously reinforced by sexual relations, but within the alliance of living together (e.g., marriage), love survives the age-related extinction of the sexual drive. It is a common experience that there is no explanation for tastes in the choice of a partner. The tie is based on the acquired drive induced lifelong fixation of an almost unlimited number of chains of ICRs that can be ecphorized any time. Nevertheless, in this case, the sea of almost irreversibly fixed engrams are interwoven with the sexual drive in a complicated manner.

The nature of cortical enhancer regulation makes it clear that everybody tries to acquire the drives best fitting to his/her brain. Nevertheless, the phenomenon described as boredom also accompanies this most sophisticated form of human acquired drives. To eliminate boredom not only the drastic change of partner, but sometimes mild changes may sufficiently activate the proper groups of cortical neurons. The common human experience, that a languishing attachment might temporarily be reinforced even by such simple changes in the boring environment like a new hairstyle, a new dress, and so on is astonishingly similar to what was shown in glass-cylinder-seeking rats in Protocols 1, 2, and 3.

*

All in all, regarding the mechanism of the dramatic changes in behavior shown in Protocols 1, 2, and 3, we uphold the opinion described in our earlier study (see Chap. 6 in Knoll 1969). Under artificial experimental conditions an active inhibitory process develops regularly in the cortical neurons involved in an acquired-drive-directed behavior that can be disinhibited in an explosion-like manner by an environmental change eliciting the orienting-searching-reflex activity, the natural mechanism that keeps the cortical neurons active. Thus, boring is an artifact, the consequence of unnatural living conditions. The phenomenon was never observed in animals living under natural conditions, but is detectable in our glass-cylinder-seeking rats, which acquired an unnatural drive under laboratory conditions. The conditions of the human beings living in developed countries are usually also far from natural ones, and their acquired-drive-directed behavior is also subject to similar changes as those shown in our glass-cylinder-seeking rats.

*

It seems reasonable to repeatedly point to the special importance of the "active state" for brains capable of acquiring drives. The healthy brain escapes from a constellation that is not determined by at least one acquired drive. The optimal condition for cortical neurons is to step out from the inborn naive state (Group 1) and start to reach a functionally higher level of organization. It seems reasonable to assume that the acquisition of chains of ICRs or new drives is the physiologically most attractive, most desired, most sought-after state for cortical neurons. Moreover, it is the optimal condition for the brain to work under the influence of a drive, i.e., to be in a state in which a group of cortical neurons are permanently maintained by their specific enhancer substance at the highest level of excitability ("active focus"). Whoever built an acquired drive into the brain of a dog, experienced the animal's extreme joy in exercising the acquired goal-seeking activity and also witnessed that the animal spares no effort to reach the goal. Humans know from their experience that they prefer to be in an active state that is pleasant, amusing, that makes them happier and more satisfied than to be in the vigilant leisure state. It is natural for humans in possession of a proper work-related drive that their preferred activity never makes them tired. Creative minds demonstrate this physiological endowment of the human brain most convincingly. Mozart wrote once to his father that to compose music is the rest for him and the inability to do so immediately tires him. Millions and millions in possession of a proper work-related drive could have written this letter.

*

Comparing human experiences with our finding that a glass-cylinder-seeking rat that performed untroubled for a long time suddenly gets bored suggests that this is a universal mechanism that can disturb any form of acquired drive-induced goal-seeking behavior. It may, for instance, disturb the flawless creative performance of an artist and induce in him or her a feeling of loss of creativity with unforeseeable consequences.

The mechanism whereby the excitability of cortical neurons substantially decreases in response to long-term tedious stimulation is linked to an unpleasant affective state of consciousness. Humans suffer from the state of boredom, and it is not by chance that the phrase "bored to death" exists. It is the physiologic ambition of the healthy human brain to stay continuously in a drive-motivated active state. Sleeping is sufficient for rest.

The fact that it is the physiological endowment of the human brain that it tries to escape from inactivity and abhors boredom is of immense practical importance. In our ceaseless effort to prevent/eliminate inactivity we always look for new stimuli that arouse the orienting-searching reflex activity. This *scourge of nature* drives the human brain to discover the world. It has always driven and will forever drive discoverers to offer themselves up voluntarily for their planned voyages. All the explorers who have died in the past after prolonged and painful misery at sea, in jungles, at the North or South Pole, or in the mountains were fully aware of the serious risk their expeditions involved. The impartial observer may sometimes harbor the feeling that the dangerous undertaking is beyond reason, but human activity depends on the acquired drive only. The most adequate answer to the question of the motivation underlying life-threatening undertakings was provided by a world-famous climber, George Leigh Mallory, who tried to conquer Mt. Everest in 1921 and 1922 without success and finally disappeared on the northwest ridge in 1924. When asked why he wanted to climb Mt. Everest, he replied "because it's there."

5 Theoretical Aspects of the Enhancer Regulation Approach

5.1
Simultaneous Coexistence of Determinants of Order and Chaos in the Human Brain: An Approach to the Origins of Science and Art

Metaphorically, every human being is born with a telencephalon that resembles a book with over 100 billion empty pages (untrained, naive cortical neurons, Group 1), and with the capacity to inscribe as much as possible in this book throughout life. In reality, the cortical enhancer regulation, the modification of the presently still unknown chemistry of the cortical neurons through learning, aiming to establish a cooperation between cortical neurons that did not know each other before, is the essence of human life.

Concluding from the analysis of learning-induced behavioral changes during the process of the fixation of a glass-cylinder-seeking drive into the brain of rats, we postulate that the chemical changes in the cortical neurons develop in three steps. Furthermore, as a consequence of this, the untrained cortical neuron (Group 1) ascends in the hierarchy and thus assumes the role of those of the Groups 2, 3, or 4. The functional significance of the cortical neurons changes accordingly. They get involved either in: (a) a chain of ECRs (Group 2), or (b) a chain of ICRs (Group 3), or (c) in an acquired drive (Group 4).

Considering humans' personal experiences, it is easy to comprehend that whenever a chain of ICRs or an acquired drive is in operation the biochemical chain of events that proceeds in this functional state is inseparable from

1. Cognitive/volitional consciousness, the perception of a psychic experience, the apperception of the self, and

2. Emotion, an affective state of consciousness, in which joy, sorrow, fear, hate, etc., are all included

This means that whenever a drive is acquired, chains of ICRs are fixed in the brain; thus, groups of telencephalic neurons change irreversibly through practice, training or exercise, as described in Sect. 3.4.2; at the same time, groups of neurons responsible for emotions are inevitably coupled to the integral whole that can be ecphorized later at any time in the future.

The long-term observation of the behavior of glass-cylinder-seeking rats convinced us that even a rat influenced by an acquired drive must be subject of a kind of psychic experience inseparable from emotions. This accords well with common experiences regarding the behavior of domesticated animals. We found convincing *experimental* evidence in support of this reasoning in a study by Phillips et al. (2003) on a peculiar form of goal-directed behavior performed by rats.

They trained rats to self-administer cocaine and paired lever-pressing to an audiovisual cue. They measured in this type of goal-directed behavior the release of dopamine from the nucleus accumbens via fast-scan cyclic voltammetry (FSCV). This is presently the most sensitive method to measure subsecond changes in the release of chemicals in the brain (Stamford and Justice, 1996). The key to FSCV is a probe $10 \mu M$ in diameter. The probe was inserted into the nucleus accumbens of the trained rats. They randomly presented the audiovisual CS in the absence of lever-pressing. Stimulation of the cortical neurons by the CS caused a rapid increase (93.9 ± 12.2 nM) in the extracellular dopamine concentration, beginning at 0.1 ± 0.0 s, peaking at 2.2 ± 0.5 s and returning to baseline 5.7 ± 0.8 s from the probe onset. But only in trained rats was the conditioned stimulation effective in producing dopamine release, *in untrained rats this stimulation was ineffective*. This is clear proof that when the chain of ICRs was fixed through training in the cortex, mesencephalic neurons also joined the integral whole as appurtenances, as adjuncts and were activated whenever the irreversibly fixed chain of ICRs was ecphorized.

We assume that whenever a drive is acquired, chains of ICRs are fixed, and also neurons responsible for emotions are coupled to the integral whole; thus, cognitive/volitional consciousness is necessarily inseparable from an affective state of consciousness.

The mechanism that binds emotions as appurtenances to any chain of ICRs is of crucial importance to interpersonal communication. For example it can be construed as a physiological basis for the enjoyment of art. Let us take Beethoven's Sonata No 106 for pianoforte, one of the immortal pieces of piano music, as an example, to illustrate from a physiological point of view: (i) the origin of the delight in music, a typical form of art without narrative function, and (ii) the origin of our ability to discriminate between two performers of this sonata who both play exact renditions of it.

When Beethoven reached his goal by trial and error and finished the composition, he strung sounds together in a sequence that never existed before and laid down the composition in an exact order of notes, making its revocation accessible to anyone with the necessary knowledge. In the brain of Beethoven the piece of music as an integral whole was also necessarily coupled to emotions in a highly complicated, never repeatable manner.

When a piano artist, driven by the acquired drive to learn the sonata for public performance, gets through the very hard training period and finally acquires the ability to properly transmit the message to the audience, the process in the brain of the artist is *essentially* similar to the changes in anyone's brain (including Beethoven's) who is capable of ecphorizing the sonata as an integral whole.

During the training procedure the artist first fixes the complicated chains of ICRs in exact order. Further hard work allows him or her to gain the ability to faultlessly ecphorize at any time the sonata as an integral whole.

The artist is reaching the goal when not only the millions of the fixed chains of ICRs can faultlessly be ecphorized in exact order, but the sonata as a whole is transmitted in an individually interpreted manner which is emotional, enthusiastic and highly characteristic of the performer.

Whenever the sonata is correctly played, the sequence of the sounds must be the same. The initiated listener who can perceive and even check the exactness of the performance is not satisfied by technical brilliance alone. He takes real delight in the indefinable, emotional side of the presentation, in the individual quality and richness of the appurtenances. Being enraptured by the performance he may even overlook some minor technical laxness. Thus, the physiology that makes *cognitive/volitional consciousness necessarily inseparable from an affective state of consciousness,* and the limitless individual differences in the appurtenances allow that the same piece of music can be presented in an emotionally endless variety.

In the brain of the audience, even in those who hear the composition for the first time, the music activates a population of cortical neurons essentially similar to those activated in the brain of the artist who transmitted the message. Due to the substantial individual differences in experiences there is of course an immense variety in the cognitive/affective reception of the performance. However, despite these differences, when a piece of music is authentically rendered by an inspired artist, the brilliant performance captures the emotions of the whole audience. It is easy to see that in the unanimous final ovation of the artist the mass-effect also comes across and makes the performance a delightful and memorable experience even for those who are less initiated in music appreciation.

*

Naive cortical neurons change perpetually from birth until death. Whenever the chemistry of cortical neurons changes through practice, training, or experience in a manner that makes them part of Group 3 or 4, this functional change is irreversible. The permanently fixed chains of ICRs or acquired drives remain simultaneously alive until death. We may be considered as much like some of the entries filled into the billions of empty pages of a telencephalic book. Each human brain is theoretically capable of everything which has been ever created by any human being because the empty book and the mechanism of inscribing into it are the same at birth. This, however, is just a potentiality and obviously unrealizable. Due to the fleeting character of life we are only able to scribble on a very small percentage of the pages, although much depends on the circumstances of life. But even this low efficiency means that a diligent human being may change, through the learning process, the functional significance of billions of cortical neurons during a long lifetime, build tens of millions of ICRs and a high number of acquired drives.

It is clear that at any given time only one of the innate or acquired drives can prevail. Nevertheless, if, for example, we put a well-trained glass-cylinder-seeking rat in a new environment, we can certainly predict that the rat will sooner or later, by trial and error, find the glass cylinder and jump on top of it. Because of the physiological nature of the cortical enhancer regulation, the exact chain of events (i.e., all the behavioral patterns that will appear until the rat finds the glass cylinder) is, however, unpredictable. The unpredictability of behavior is most pronounced in humans as they possess the most sophisticated brain.

We postulate that cortical neurons belonging to Group 3 or 4 continuously synthesize their specific enhancer substance within their capacity. This means that even in the vigilant resting state (leisure), in the absence of a dominant drive, as well as in the non-vigilant resting state (sleeping), the cortical neurons representing the totality of the already fixed ICRs and acquired drives are permanently under the influence of their specific enhancer substance. In addition, as learning means that cortical neurons acquire the ability to be activated through alien enhancer substances too (see Sect. 3.4.2), they may also be continuously influenced by enhancer substances other than their own. Although the level of this permanent, undulating activation remains low, it is unpredictable as to when any group of cortical neurons will be influenced by enhancer substances on the level already inseparable from conscious perception. Thus, as the totality of the cortical neurons belonging to Group 3 or 4 works continuously on an unconscious level, there is a steadily operating, chaotic background noise in the human telencephalon. Although in the active state ("fight or flight" behavior, goal-seeking), when the actually dominant drive determines the rational goal to be reached, and the background noise is suppressed, it can never cease to exist. But it never endangers the function of the actually dominant innate or acquired drive. From this situation it follows that *rational brain activity is necessarily amalgamated with irrational brain activity, and we live through every moment of our life experiencing the totality of order and chaos in our brain.*

Without even trying to touch upon the peculiar world of psychoanalysis, there is one aspect in Freud's life-work, maybe the most important one from a physiological point of view, that seems to me to be unquestionably in accordance with the cortical enhancer regulation approach.

In my view, Freud intuitively realized not only the physiological situation described above, but also the facts that: (a) whenever a chain of ICRs is fixed in the human brain, the proper cortical neurons remain for life, on an unconscious level, constantly active, and (b) by the aid of a proper method, even a chain of ICRs that was never ecphorized after fixation can be activated to the level needed for conscious perception at any later point in time. The recollection of any chain of ICRs is necessarily inseparable from an affective state of consciousness, due to emotions coupled as appurtenances to cortical neurons when these neurons learned to cooperate with each other.

Freud developed empirically proper methods for ecphorizing forgotten chains of ICRs in humans, decades after their fixation. Because the physiological basis of this method was unknown, a special terminology was developed to describe the findings. The frequent misinterpretations and pseudo-scientific deductions made from the results produced by different methods of recalling "forgotten" experiences in humans do not detract from the merit of the basic technique originally introduced by Freud.

*

We may summarize our view about the requirements that determine the cortical activity of species capable of acquiring drives as follows:

1. Past experiences are irreversibly fixed in neurons belonging to Group 3 and 4 that learned to cooperate with each other and constitute an integral whole.

2. Proper stimulation of the cooperating neurons as an integral whole allows the ecphorization of the fixed information at any later point in time. This is inseparable from conscious perception; thus, the past experience is vividly relived in a cognitive and affective manner.

3. Even though the individual is primarily focused on reaching the goal represented by a dominant drive during its operation (rational activity), the ability to simultaneously and consciously revive past experiences that are outside the limits of the actually operating dominant drive (irrational activity) is a natural endowment of the brain.

The working model that every moment of our life is determined by the actual state of the amalgamation of order and chaos in our brain allows a new approach to the interpretation of attention. Psychology defines attention as the concentration of the mind on a single object or thought. According to our approach, attention is a constellation in the brain occurring when a number of cortical neurons belonging to Group 4 produce their specific enhancer substance in the optimal concentration. These neurons are in a state of highest-level excitability ("active focus"), thus suppressing chaos in the brain as much as possible, and the organism concentrates on reaching the goal determined by the active focus.

Amalgamation of order and chaos in the brain can be detected easily in humans, as they can give account of their psychic experiences. Penfield, in his paper entitled "The permanent record of the stream of consciousness", was the first to furnish clear-cut *experimental* evidence for the amalgamation of order and chaos in the human brain. He employed gentle electrical stimulation to circumscribed cortical areas in the exposed brain of fully conscious patients. In response to stimulation the subjects vividly relived past experiences while at the same time being aware that they were in an operating room (Penfield 1955).

*

It is well-known that when Pavlov, the first scientist who dared to approach behavior as an objectively analyzable physiological function of the cortex, tried to understand the essence of changes in cortical neurons in the course of the acquisition of an ECR, he was compelled to carefully isolate his animal from the outside world. The slightest alien noise that distracted attention compromised the success of the experiment. The reason for his difficulties now seems to have become clear.

Learning means the development of mutual attraction between cortical neurons that had no previous contact with each other. The simplest form of it is the development of a reversible, transient attraction that is easily lost. This is the transition of a naive cortical neuron into one that participates in an ECR. Pavlov restricted himself to following only this first step in the learning process, the transition of a cortical neuron from Group 1 to Group 2. During the operation of a drive, when the excitability of the cortical neurons is enhanced and chaos is suppressed, the cortex rapidly acquires new chains of ECRs as this process is the *sine qua non* to the rapid adaptation of the subject to an ever-changing outside world (see Chap. 4 in Knoll 1969, for review). It is, however, extremely difficult to build an ECR in the absence of a dominant drive, since the ever active chaos in the brain efficiently suppresses the fixation of any new, temporary connection. Pavlov was therefore compelled to use special techniques, such as properly sound- and light-proof chambers, to isolate his animal as much as possible from the chaotic influences of the outer and inner world.

*

The naive cortical neuron is born with the ability to perceive one of the senses: color, light, pain, sound, smell, taste, touch. As all neurons, a cortical neuron also responds to its specific stimulus with an evoked potential. The appearance of this change within 0.015 s gives information about the prompt perception of the stimulation. The essential physiological function of a cortical neuron is, however, to get involved in hitherto nonexistent cooperation with other neurons (learning). Although the neurochemistry of the learning process is still unknown, we can accurately measure its consequences. We can follow the sequence of behavioral modification induced by the learning process. We can define three essential forms of behavioral modification that differ in kind. The reversible form of this attraction is the ECR. Its physiological significance is to secure in goal-seeking behavior the rapid adaptation to new conditions. In contrast, the process of learning leading to a stable, irreversible attraction between cortical neurons appearing either as an ICR or as an acquired drive results in functions designed for lifelong operation.

According to our working hypothesis (see Sect. 3.4.2), a chain of ICRs is ecphorized as follows. The proper stimulus activates the first member of the chain. Activation means that a higher amount of the neuron-specific enhancer substance is synthesized and the excitability of a mass of cortical neurons is enhanced. The members of the chain are thereafter explosively activated in the sequence in which they were fixed during the learning process. Activation of the chain of ICRs as an integral whole is inseparable from conscious perception and from the proper affective state of consciousness.

The performance of the most famous artists, musicians, and actors illustrates the immense capacity of the human brain to fix long chains of ICRs and ecphorize them whenever needed as an entity. We may take as an example the performance of a talented pianist who can play faultlessly for hours, even by heart, the most difficult compositions in succession. An extremely well-operating acquired drive in the brain of the pianist was the main condition for getting through the exhausting learning process required to properly fix those tens or hundreds of thousands or even millions of very long chains of ICRs in the cortex and ecphorize them whenever needed in a faultless manner. It is the ease and high speed in fixing chains of ICRs that make the human brain qualitatively different from the brain of animal species already capable of acquiring drives.

The inborn enhancer regulation operating in the brain stem is based on a couple of innately fixed enhancer substances. Although only two of them, PEA and tryptamine, have been investigated so far, there is good reason to predict that many other potent endogenous compounds exist and play a role in the enhancer regulation. In addition, we already possess synthetic mesencephalic enhancer substances [(−)-deprenyl and (−)-BPAP].

In striking contrast we have no similar experimental tools to influence the learning performance of the cortical neurons. As was shown in Sect. 3.2.2.3 (Table 3.2), none of the 10 applied concentrations (ranging from 0.000001–10 mg/kg) of (−)-BPAP, the most potent mesencephalic enhancer substance presently known, enhanced the activity of the cortical neurons.

Because of the theoretically immense variability of the cortical enhancer regulation, any trial to develop a compound that will reasonably stimulate learning in general seems to be a lost enterprise from the physiological point of view. It is very likely that the natural method of modifying behavior through experience, training, or practice remains not only the most effective, but will presumably forever remain the only viable way to change the performance of the cortical neurons in species capable of acquiring drives. For this reason, everything depends and will probably always depend on teaching, learning, education.

The relationship between cortical neuronal activity and conscious perception can be studied in humans only. It is reasonable to assume that in the process whereby cortical neurons learn to cooperate with each other and the modification of behavior is fixed for a lifetime, a complicated, slowly developing new chemical constellation is necessary for this change. It seems obvious that to bring the masses of cooperating cortical neurons to the level of excitability needed to ecphorize the fixed information as an entity must be a substantially longer process than the 0.015 s needed for the appearance of an evoked potential following stimulation of a cortical neuron with its specific stimulus. As discussed earlier (see Sect. 3.4.2), this is true. It was shown in a human study that 0.5 s was needed for the conscious perception of a cutaneous stimulation (Libet 1973). The cortical enhancer regulation approach gives, for the first time, a reasonable explanation as to why a 33-time-longer latency period (0.5 s) was required in the Libet experiment for the conscious perception of a cutaneous stimulation that induces an evoked potential within 0.015 s.

*

As an innate drive establishes an acquired drive, we may assert that nothing exists in the brain without a rational origin. Furthermore, we cannot forget that at any point in time it is just by chance which of the innate or acquired drives will be the dominant one. The operation of a dominant drive and the related chains of ICRs means that a limited number of cortical neurons are at their peak of activity, while all the billions of cells representing chains of ICRs and other drives work at the customary unconscious, undulating, low-activity level, constituting the background noise in the brain. Thus, a cortical neuron that changed its functional significance through learning and already belongs to Group 4 is a Janus-faced cell. It rises any time it is needed into the limelight and then operates as a dominant cell; once the need is over it, it sinks into obscurity and drowns again in the sea of cortical neurons maintaining the chaos in the human brain.

I conjecture that each side of the amalgamation of order and chaos, of the simultaneity of the rational and irrational in the human brain is of equal physiological importance for human life. We owe to the rational brain (order), the creator of science, our ability to survive in the real world, and we owe to the irrational brain (chaos), the mother of art, our ability to escape from reality. A minority born with an unusually high sensitivity to realize the chaos in their brain will aspire to find their place in art, whereas another minority endowed with a diametrically opposed sensitivity will strive to be productive in science/technology. The overwhelming majority, being devoid of special affinity in either direction, is saved from the irrepressible drives to create something and live in obscurity.

*

According to the approach presented in this study, the self is at every moment the sum of the cortical neurons that belong to Groups 3 and 4. The nature of the acquired drives, the hierarchy of the importance of the goals they represent, determine the life of the individual. As the perception of the self is also inseparable from an affective state of consciousness, satisfaction is necessarily proportional to the success in reaching the goals represented by the innate and fixed acquired drives.

To make a living everybody has to find a place in society. One can choose only from the opportunities given and is compelled to accept harsh realities. Life in the community is usually designed in accordance with the faculties of the average individual who is free of a passionate, irrepressible vocational drive, and is always ready, without noteworthy conflicts, to remain within the bounds of given possibilities. Anybody who tries to go to any great length to reach a "higher" goal may provoke a conflict. Creative minds are especially endangered.

The relation of geniuses to the society in which they live is for various reasons, but at all times, plagued with serious, sometimes fatal conflicts. To illustrate this, we need only recall, from among the thousands of well-known examples, a classic one, that of the unfortunate lot of Mozart. He was born and raised in the still strictly aristocratic Austria, where a musician could make his living only in a court. The traditional spirit of the court did not differentiate between servants, for example, between a cook and a musician. Mozart, one of the most creative artists in the history of mankind, was obviously aware of the immortal value of his music and wanted to be an independent artist in a time when the aristocratic spirit in his native country made such effort impossible. Thus, a fatal conflict was created. He died impoverished, forsaken by all, at the early age of 35 in 1791.

It is amazing that Mozart died in the period when Paris rose in revolt, announcing, *inter alia,* the program to abolish the privileges of the nobility of birth. As this objective was viable, with time already ripe for this change, the French Revolution was victorious from point of view of Liberty. The traditional aristocratic Weltanschauung *slowly* faded away. The immense strength of the privileges of the nobility of olden times already belongs to the relics of the past.

In contrast, the program of the Revolution to abolish forever the privileges relating to property could not be accomplished. This objective was condemned to fail for a simple reason. To live in a society is a chance to surpass qualitatively the performance of its individuals. The essential aim is to be more productive. The main function of the society is its distributional activity. The condition *sine qua non* for the productivity of a community is to give everybody free scope for his abilities and to bestow special care on talent. It belongs to the physiology of the brain that there is an enormous variation in abilities and only a minority has talent. Thus, the contradiction between Liberty and Equality is a natural endowment. To minimize the consequences of this natural contradiction and *keep simultaneously productivity and the inner harmony of society at an optimum level* has always been and will always remain the most significant challenge for the ruling élite. Unsuccessful experiments to create by force a society free from property-based privileges (e.g., the Soviet Union) proved that the overemphasis on Equality restricts the self-assertion of talents, cripples productivity, and makes such societies uncompetitive with those free of these anomalies.

*

Art, being independent from external reality, is a logical product of chaos. The ability of the human brain to create an "other world" is unique. "Reason is the enemy of art" (James Ensor). Due to the chaotic function of the brain, humans created the world of myths, a universe that is in striking contrast to the rational world. In this irrational universe, we are unrestricted, free, and happy. "Day-dreaming is the Sunday of thinking" (Amiel). To compensate for the harsh reality of their lives, humans badly need the boons of the self-created, nonexistent universe.

Religions with their highly efficient myths, with their Holy Scriptures, with the creation of Gods, and later of one God, belong to the highest level of artwork ever produced. They are the most fascinating and hitherto most influential products of the art-creating human brain. They played an obviously decisive role in the fabrication and amalgamation of the ancient human communities prior to the development of a significant level of science and technology. Religion is still an absolute necessity for the majority of human beings. The reason is clear. The creative scientists or artists have their acquired drives that totally occupy their lives. Innate drives cannot fully satisfy humans who possess a telencephalon with an immense capacity to acquire drives. Goethe clearly understood the problem: "He who possesses art and science has religion; he who does not possess them, needs religion."

It is a physiological necessity that something like a "Panem et circenses" approach keeps human society running. To supply the masses with proper acquired drives that add color to their otherwise boring lives was always the main concern of the ruling élite. With the passing of time, the technology of the "Panem et circenses" approach necessarily changed a lot. The essence, however, remained unchanged. Religion and public amusement were the greatest inventions of the past to satisfy the physiological need of the overwhelming majority of the society, those whose minds were not totally occupied by a vocational drive.

Both nature and society put serious restraints on everyone. Survival in the real world demands cautious self-restraint, and first of all, laborious creation of new knowledge about the physical world, natural laws, and society with significant practical consequences. Such knowledge can only be obtained by tedious observation and a never-ending testing of facts. It is obvious that science and technology, the highest level of human adaptation to the realities of life, is rooted in the almost infinite human ability to acquire drives. Humans need their rational brain to cope with the blunt facts of life and the irrational to forget the real world.

As science is the product of rational brain activity, a scientific work, an analysis of a piece of reality, has essentially the same message for all who are blessed with the necessary prerequisite knowledge. In contrast to science, it is not the objective of fine art, literature, or music to describe or analyze objectively existing reality. On the contrary, real art deals with a nonexistent world that is indefinable in any material sense, with a universe created in the human cortex. The message of an artwork is therefore ambiguous, its effect is in essence emotionally based. Everyone enters in his own special way into the spirit of an artwork. This is also true for the Holy Scriptures, art products of the human brain that played the most important role in the spiritual organization of life in the rising phase of human society and served the maintenance and development of that society.

The entire, invented "other-world" that art (chaos in the human cortex) has produced, including God, Satan, angels, demons, and so forth, all the creatures that people the imagined, nonexistent, human-brain-created universe, is rooted of course in the existing reality around us. For example, humans abhor bats, and the Evil One soars on bat wings in the medieval paintings; or the transubstantiated bloodsucking bat appears in the mythical figure of the vampire, and so on. But, just as with the observation of an acquired-drive-induced behavioral performance we are unable to detect the innate drive that made its acquisition possible, it is sometimes hard to recognize the transubstantiation of the blunt reality into some complicated creatures that art created within a nonexistent world of myths. But this is, after all, not essential. We all badly need the emotions aroused by the nonexistent world as created by the human cortex, and we also need a satisfying interpretation of the human psyche.

The concrete biochemistry of cortical enhancer regulation, the puzzle of the amalgamation of order and chaos in the human brain is still unsolved. As usual, attractive descriptions remedy such deficiencies. Freud's work was the hitherto most influential description of the coexistence of rational and irrational, conscious and unconscious, in the human brain. It is no accident that his fascinating writings, enjoyable readings closer to fiction than science, have always deeply influenced many artists. Freud also provided a delightful interpretation of dreams at the boundary between art and science. It was probably the pseudo-scientific nature of his brilliant books that precluded him from receiving the well-deserved Nobel Prize for literature.

As the background noise in the brain is never interrupted and can even be more accentuated during sleep than in the vigilant resting state, the dream-world, the classic example of a man-created universe, always gave inspiration to art. Surrealism is the best example of this inspiration. There can be little doubt that already in the past – from the Greek masters through Shakespeare – human behavior was described in all its details with a hardly surpassable quality. Nevertheless, these are only wonderful descriptions. From the scientific point of view they are irrelevant. The still unsolved scientific problems that await their disclosure: the chemistry of cortical enhancer regulation and the natural law that determines the operation of the brain and its self and is responsible for the immense variability of human activities represent the ultimate explanation for human behavior.

Human society, the maintenance of which has always required the proper manipulation of the brains of its members, is still in development. It seeks its final equilibrium: This will be the state reached when the modification of behavior induced by the triad of family/school/society is based from birth until death on the exact knowledge of the natural laws that keep the brain and its self going.

The French Revolution was the first attempt to change the traditional manipulation of the human brain. It was also clear that the development of science is the only way to reach this aim. Diderot's maxim: "It is very important not to mistake hemlock for parsley, but to believe or not believe in God is not important at all" shows that the pioneers of Enlightenment realized 200 years ago that the basic spirit of education needs a qualitative change. This is still a pressing necessity.

With the disestablishment of the Church, the French Revolution took the first necessary step in the right direction. This measure was later successfully implemented in many countries that developed thereafter more rapidly, but even in these countries the traditional spirit of education, based on the belief in supernatural forces, did not change too much. Only further development of science, the knowledge of the real mechanism of action of the human cortex may, in the long run, change the situation in a revolutionary way.

It is remarkable that some artists comprehended intuitively the ambiguity of our brain work. Alfred Kubin wrote in his novel *Die andere Seite* (first published in 1909): "For me it has gradually become a certainty that the human being consists of a puzzling amalgamation of two anonymous entities Chaos and Self." Niki de Saint Phalle asked in one of her paintings, entitled "Order and Chaos" (1994, Collection MAMOC, Nice), the following questions: "Order/chaos (I am chaos)? Which came first? Do they exist side by side?" I firmly hope that the physiology of order and chaos in the human brain answers these questions.

Poets capture the essence of order and chaos in the brain in a few lines. They somehow really describe the indescribable. For example, in a very short, unusual masterpiece ("Milyen volt"), the Hungarian poet, Gyula Juhász, tears us away from reality and we follow he who *sees* the ripe wheat ears and *feels* the blondness of his quondam love; *sees* the shattering blue of the clouds and *feels* her blue eyes; *hears* the vernal breeze and *feels* the warmth of her voice.

5.1 Coexistence of Order and Chaos in the Human Brain

Examples illustrating that chaos in the brain is the mother of art can be quoted *ad infinitum*.

*

The requirement that art should correspond to Nature was the natural view from the very beginning until the end of the 19th century. Leonardo, the enigmatic genius, who remained the unsurpassed and key figure in the development of European art, represented this approach most convincingly. He admired Nature and his thirst for knowledge and discovery was insatiable. Everything had to be both natural and rational in his paintings. He painstakingly studied anatomy, mechanics, hydraulics, etc. For him to paint his *Last supper* was as natural and important as the construction of a device for raising a mast, or to write an essay on the reason why distant mountains and objects appear blue. All in all, the approach that art should correspond to Nature remained the canon of good taste for centuries. The consequences of the Enlightenment, the revolutionary breakthroughs in science/technology during the 19th century, the modernization of life at a pace never seen before changed necessarily and unavoidably the content of education, the millions of chains of ICRs fixed in young brains, thus transforming society rapidly and deeply and, with the passing of time, disrupting traditionally developed canons.

To exemplify this change, we may briefly follow the developments by which the canons of fine art changed during recent centuries. Although efforts in religious art to somehow express the envisioned ethereal "other world" was a new element that had already reached a high point in the 16th century, in the painting of El-Greco, with a unique ecstatic, spiritual, mystic distortion of the figures, the rules of traditional painting remained unchanged until the 1870s. Impressionism, considering light and the exchange of colored reflexes as the unifying element of a picture, changed this. Nevertheless, even this breakthrough had nothing to do with the conscious revolution leading to the final state, the total departure of fine art from the depiction of reality. The very beginnings of the trend that led to the present state can be detected in a number of important artworks created at the end of the 19th century.

The start of the *conscious* breach with the view that the artist's task is to depict reality can be beautifully traced to the work of Cézanne, one of the founding fathers of modern art. The description of the "Still life with a plaster cast" in the Courtauld Gallery in London is a good example for illustrating the very beginning of the changes leading to the explosion in fine art during the early decades of the 20th century.

The art historian describes this painting as follows: "This is one of Cézanne's most complex late still-lifes. Beyond the foreground table with the plaster Cupid, the space and the arrangement of objects become highly ambiguous. The green apple on the floor itself in the far corner seems too large and the floor itself appears tilted. A painting is propped up against the wall at the left but its painted blue drapery merges with that of Cézanne's still-life. Cézanne may have been using these paradoxes to stress the artificiality of the still-life and perhaps even of the art of painting itself."

In the same year that Cézanne died (1906), the "Fauve" era culminated with Matisse's masterpiece "The joy of life." One year later, in 1907, Picasso initiated cubism with his "Les Demoiselles d'Avignon." It is easy to detect in the works of Picasso, Miró, Klee and others that in this transition stage of modern art the revolutionary artists were fascinated by the magical power of prehistoric art and the peculiar spirit of primitive art.

In retrospect we may say that the revolutionary developments in fine art toward abstraction followed a series of logical steps. After the ideologically supported beginnings traceable to the lifework of Cézanne, the natural next step was the fractionation of the objects into pieces. Picasso was undoubtedly the spiritual initiator of the new trend, but the unique experimental period in the history of modern art was related to the legendary collaboration of Picasso with Braque in the years from 1911 until 1914. As a matter of fact it was the cubist revolution that ultimately terminated the "narrative" period of fine art.

With the fragmentation of the object and the proper coloring of geometricized shapes, hitherto unseen rhythmic compositions were created, capable of exciting millions of cortical neurons in the brain of the initiated viewer inseparable from feelings of harmony and satisfaction similar to those elicited by music. The final step of course was the complete omission of objects and the deconstruction of compositions that act only by the power of colors, as music does with sounds.

An obviously decisive turning point was therefore Kandinsky's approach. He was the artist who proclaimed in his book *Über das Geistige in der Kunst* (Concerning the spiritual in art), published in 1911, the theory of the departure of art from the objective world. Kandinsky was once asked in the late 1930s how he arrived at his abstraction in painting. He pointed to the essence of the problem: "I envied musicians who could make art without *narrating* anything realistic. Color seemed to me just as expressive and as powerful as sound." The exhibition "Aux origines de l'abstraction (1800–1914)" in the Musée d'Orsay (Paris, 2003) was probably the most instructive résumé yet of the crucially important developments in modern art, tracing back the changes to their origin.

There is no doubt that the artists turning away from the traditional, exact reproduction of the world as perceived by the senses was of crucial importance. For example, Jawlensky, the painter who for years worked closely with his friend Kandinsky, wrote in his memoirs that it was in 1905 that he first understood not to paint what he *saw* but what he *felt*. As a matter of fact, this was the leitmotif of Kandinsky's epoch-making book. It was probably not by chance that *The Art of Spiritual Harmony* was the title of the first English language edition of Kandinsky's book in 1914.

From the point of view of this study, Jawlensky is an outstanding example for showing the enormous power of an acquired drive in humans. He fell seriously ill with an extremely painful form of arthritis that practically made him unable to work. In the final stage of his illness he worked simultaneously on eight small pieces of cardboard that were fixed near each other like an iconostasis and produced his last series of variations on the human face, entitled "Meditations". He produced about 1,700 (!) such pieces in periods when the continuous suffering let up a little bit and allowed him to work.

From the physiological point of view the teaching of the life of Jawlensky is replicated in the life of Frida Kahlo, a painter with a touch of genius. She sustained a bad accident in her youth that seriously injured her hip bone. As a consequence of the severe injury she not only suffered all her life, but was also unable to have children. Her work was imbued with the frustration caused by three unsuccessful attempts to give birth to a child. She was confined to bed during the last three years of her life, was subject to a series of surgical interventions, and one of her legs was finally amputated. Even in this almost continuously painful last period of her life she never ceased to work. Once, near the end of her life she wrote in her diary that she has three wishes: to paint, paint and paint. She finished her last beautiful fruit still-life eight days before she died.

In the 20th century, due to the revolutionary spirit of modern art, even architecture changed dramatically. For example, with the "Falling water" house, this still unsurpassable masterpiece, Frank Lloyd Wright succeeded in expressing the indescribable feeling of the man who seeks repose in the beauties of nature by creating a harmony never seen before between a forest, stream, and rock, and all elements of the structure. He was 72 years old when he finished this masterpiece and 88 when he started to create the Guggenheim Museum in New York, the most revolutionary museum building of his times. He died in 1959 with a creative mind as productive as ever.

Creative minds give the best proof that acquired drives determine human life, and they are also the best examples for deeply penetrating into essential aspects of acquired-drive-dependent behaviors.

Acquired drives work on the highest activity level attainable in creative minds. The "active focus", the cortical representation of the acquired drive, a population of cortical neurons which belong to Group 4, synthesize, when activated, their specific enhancer substance at the optimum level. This is the situation in objective terms. Subjectively it means the imagination of the goal that the individual expends his entire effort to attain.. Nevertheless, the goal is, even for a genius, not necessarily always reachable. Again, Leonardo da Vinci, one of the greatest creative minds in history, is probably the best example for illustrating this paradox. Leonardo – whom Nietzsche mentioned as a fine example of "those marvellously incomprehensible, and inexplicable beings, those enigmatic men, predestined for conquering and circumventing others" –

was never satisfied with what he did. He often mentioned that *his hand could not match his mind* and completed only half a dozen of paintings.

*

It seems to me that without taking into consideration the simultaneity of order and chaos in the brain, the operation of human society is incomprehensible. Language, the most important tool in interpersonal communication, must be clear, fully understandable in service of rational activity. This is obviously the condition *sine qua non* for adaptation to reality, thus, for survival. For the chaos in our brain, however, the "meeting of an umbrella with a sewing-machine on the dissection-table" is not only a natural event, but even a source of delight. It is somehow a touch of chaos when in conversation that serves a rational, goal-directed activity we use words without definable meaning, like thingumajig, gadget, jigger. The role of these words is exactly to be devoid of concrete meaning. In times past, whenever, in spite of all efforts, a rational, goal-directed activity was unsuccessful, for example, when drought endangered the crop, a present, an expiatory sacrifice was offered to seek the help of God(s). This was the method how chaos in the human brain has kept hope alive and helped us to survive. This vital mechanism is still in operation worldwide. Billions seek aid in their conflicts in the church, from a psychologist, and/or from the different common forms of ersatz.

The most essential forms of behavioral functions clearly displayed on the borderline between chaos and order in the human brain are the classic phenomena of the behavior of masses. The most notorious, precisely documented examples, which clearly demonstrate the extremely dangerous character of this phenomenon in human history, are the mass-meetings in Nazi Germany. Especially their role in the manipulation of the masses in a still peaceful Europe, in the developing period of the Nazi system, before the outbreak of World War II, is remarkable. Hitler was capable of bringing the hundreds of thousands of participants at these meetings into a state in which the reasonable brain control of individuals completely disappeared and the crowd behaved as a homogeneous entity. For the efficient preparation of the crowd Hitler introduced a perfectly executed parade under arms. Such a parade was the demonstration that tens of thousands of human beings in uniform can lose, given proper training, their individuality, and behave as robots.

The natural behavior of a crowd at the borderline between chaos and order in the human brain, as experienced at sport-events, popular concerts, political meetings of extremists, and so, is always a terrifying experience for a man of poise. Its essentially dangerous, threatening nature is obvious. The history of wars prove the main role of this brain mechanism in degrading human beings into faceless, robotic members of a misused mass.

Human society, still in the process of maturity, will sooner or later arrive at its fully developed state when education will be based on exact knowledge of the mechanisms of cortical functions inseparable from conscious experience.

Then the home/school/society triad will educate the members of the community to understand that simultaneity of order and chaos in their brain is the physiological reality that determines human activity. From this time onwards it will be possible for everybody to consciously determine which acquired drives fit optimally to their natural endowments.

For the time being those who were lucky enough to acquire the best fitting drives in due time, in the early uphill period of life, had fair chances for success and happiness. In contrast, those who for any reason missed this opportunity remain frustrated and will look for "ersatz". I think order and chaos in our brain are of equal importance. *Without the ability to adapt ourselves to the concrete, we would not be able to live; without the art which allows detachment from the concrete and flights into the infinite, it would not be worth living.*

5.2
The Timeliness of the Conception of the Enlightenment: *Sapere Ande!* (Dare to Go Independently!)

The ability to fix ICRs and acquire drives made the development of social life possible. These are the brain mechanisms through which individuals influence each other's behavior and learn thereby to work together toward a common goal. The evolution of a brain with the capacity to fix ICRs and acquire drives was obviously the last revolutionary step in the development of life on earth. Thus organisms having a brain the operation of which was inseparable from some form of cognitive/volitional and affective states of consciousness appeared.

The existence of a variety of animal species with extremely restricted abilities to fix ICRs and acquire drives marks Nature's long path of experimentation with the brain. The end result of this process was the most perfect variety. About 40,000 years ago, the *Homo sapiens sapiens* appeared and the progressive development of human society, capable of consciously changing its essential conditions, began.

The limitless capacity of the human cortex to fix ICRs and acquire drives – in conjunction with the development of language – made possible an unmatched interpersonal communication. This unique facility made the cognitive/volitional and affective states of consciousness of the human brain and, as a consequence of it, human social life, unparalleled. Because of the lack of similar developments in animals, there is no way to obtain direct evidence about the nature of their psychic experiences. Nevertheless, observation of the goal-seeking behavior of properly trained monkeys, dogs, horses, and so on furnishes convincing indirect evidence for the assumption that the operation of ICRs and acquired drives even in animals is inseparable from an archetype of consciousness.

*

From the moment of its inception, the myths-based human community has undergone a progressive transformation. It still seeks its final state of equilibrium: a society, the organization of which is fully based on rationality, on the knowledge of the natural laws that keep the brain and its self operating.

All living beings with the ability to fix ICRs and acquire drives are capable of developing some form of community life, the essence of which is the development of a hierarchy in order to regulate the member's access to the commonly seized, acquired, or produced goods. The human community necessarily developed the most sophisticated hierarchy.

For thousands of years a human's place in society was already fixed at birth. Although the circumstances of human communities has changed a lot, it is still just a minority of the billions of humans living today who can significantly change their place of birth in the hierarchy.

Since the operation of cortical neurons belonging to Groups 3 or 4 is inseparable not only from cognitive/volitional perception, but also from an affective state of consciousness, it is a natural law that the individual strives for pleasant goals and tries to avoid unpleasant operations. Accordingly, the community manipulates its subjects with a sophisticated system of time-honored rewards and punishments. The creation of the myth of an *other world* was the most influential invention of the past that rendered the maintenance of peaceful order in human society possible. According to this myth, the soul is immortal. The short, transient earthly period of life is therefore of minor importance. A human whose earthly existence was honest reaps his reward in the other world, while the evil-hearted receives severe punishment.

First of all, this ingenious myth that made the believer and his beloved ones immortal helped man cope with his natural fear of death. On the other hand, the myth presented the believer with the illusion that to be rich or poor during one's very short life here on earth is insignificant. This helped the majority to tolerate the privileges related to property.

For centuries preachings and mainly paintings in the church ensured the fixation of the proper chains of ICRs and the acquisition of the drive in the brain of the illiterate masses that maintained the belief of salvation in heaven and the fear of the monstrosity of hell. The need for this belief is for the majority of people so essential that despite progressive changes in human society, the myth still plays a dominant role in the life of billions living today.

It is, however, self-explanatory that the human brain can only discover cognizable reality, understand natural laws, and make a use of this knowledge, but cannot create a world. In the center of the *Mystic Lamb*, the famous triptych painted by the van Eyck brothers before 1432, God the Almighty is a regally dressed, charming and attractive man who holds the insignia of power in his hands and has an admirable crown at his feet. The viewer is still shocked by the apocalyptic visions of Hieronymus Bosch. We meet in his paintings thousands and thousands of nonexistent creatures and objects.

5.2 The Timeliness of the Conception of the Enlightenment: *Sapere Ande!* 131

What we in fact see are just complicated variants of truly existing living beings and objects. Arcimboldo, the Roman Imperial Court painter of the 17th century, who created the most remarkable fantasy pictures, constructed from flowers, fruits, animals, and existing objects, his admired, unique heads, personifying the four seasons, the elements: air, fire, earth, water, and so on. But even art that does not narrate anything realistic, music or abstract painting or sculpture, is based on existing sounds, colors, and geometrical forms, respectively.

Whatever humans, influenced by an acquired drive, plan, devise, contemplate, and then create, results from: (a) the use of earlier fixed knowledge (ICRs) in the brain of the creator, and (b) the fixation of a series of newly acquired, proper ICRs needed to reach the goal represented by the active focus, the actually operating acquired drive. Humans obviously cannot change natural laws, but, by discovering their mechanisms of action they learn to make use of this knowledge and can even land on the moon.

*

For the mammalian brain the most important period of life lasts from weaning until sexual maturity. At the time of weaning fully developed enhancer regulation in the brain stem starts working on a higher activity level. This state lasts until sexual maturity is reached (Knoll and Miklya, 1995). For details see Sect. 3.5.1.1.

Under natural conditions living beings accumulate practically all the necessary information needed for survival during this period. Most of the ICRs and acquired drives that will primarily determine the life of a human being are also primarily fixed in: (a) this most sensitive, uphill period of life, or (b) in the early phase of the postdevelopmental period, prior to the arrival of an already advanced stage of the progressive age-related functional decline of the mesencephalic enhancer regulation. As are all physiological functions, the ability of the brain to build ICRs and acquire drives is also subject to aging. Accordingly, humans remember better the engrams fixed in their youth and forget even simpler recent acquisitions.

Because the enhancer regulation works on a higher activity level in the developmental phase of life, the younger the individual, the easier it is to manipulate the brain. Just as humans transform wild animals into obedient servants by manipulating their brain to acquire proper drives, the home/school/society triad manipulates humans from birth until death to create and maintain their compliance with the requirements of the community.

Considering that the Self is at each and every moment of life the sum of its fixed ICRs and acquired drives, it is self-explanatory that individuals living together in the same community and raised in the same cultural environment will in many aspects be very similar to each other. This is the physiological basis of the well-known and – from a practical point of view – decisively important, "national" characteristics, described as specifically English, Japanese, Russian, French, German, Spanish, Hungarian, and so on. On the other hand, it is again self-explanatory that each micro- and macro-community has in every period of its existence its own characteristic problems, which are unknown and sometimes even incomprehensible to outsiders. This is primarily due to specific living conditions and to the eternal fight of the local ruling élite for the redistribution of the leading positions in the hierarchy that exclusively occupies the attention of the members of this community.

It is the fundamental law of human society that the applied educational methods have been transmitted from one generation to the other and have been hitherto imbued with the "cultural heritage", the traditional myths known to efficiently manipulate the naive human brain according to expectation. The history of mankind is eloquent testimony that due to the almost never-failing, time-honored methods, the overwhelming majority of the individuals behave as expected. People are not aware of the manipulability of their brain. They act in the firm belief that they possess free will. This belief is unfortunately, at least for the time being and with the exception of a minority, a delusion.

*

The self of the overwhelming majority of the members of the circa 5,000 ethnic groups that populate the world today, and of the circa 200 of them that are acknowledged as states, is perpetually manipulated by the home/school/society triad. The majority behaves accordingly. They love their predetermined "friends", hate their suggested "enemies", believe that to kill the "enemy" in war is a heroic deed, and so on, and on.

The tohubohu in which the 6.5 billion humans now live together on earth is due to the heterogeneity of myths that shaped their lives during previous millennia. It is also due to the natural chaos of the myths-directed era of human society that whatever the right hand built, the left hand demolished with the passing of time. Examples: Almost none of the original works of the greatest Greek masters survived. Lysippos, one of the greatest Greek sculptors supposedly created 1,500 bronze statues, but his work is known from copies only. Leonardo envisioned and brought to a full size model an equestrian monument for Lodovico il Moro on a scale (twice as high as Verrocchio's Colleoni monument in Venice) and quality which have remained unmatched to this day. The model, a marvel of the Italian world, was displayed in Milan in 1493 and won him recognition he had long sought. Lodovico Sforza collected the vast weight of bronze required for casting this giant equestrian statue, but it never did come about because the metal was ultimately used to make a cannon, and the model itself was demolished soon after the fall of Lodovico il Moro (1499) – and so forth and so on. On the other hand, many of luckily surviving remnants of the Wonders of the World, like parts of the Acropolis of Athens, were removed from their original location and put in museums; only a future

5.2 The Timeliness of the Conception of the Enlightenment: *Sapere Ande!*

reason-directed era of human history could ever bring them back to their place of origin.

The ultimately unavoidable transition of human society from its myths-directed era into the rationally directed one will necessarily terminate the presently existing tohubohu and lead to a reasonably and harmoniously operating global human world. It is not by chance that after the cataclysm of World War II the undeviating initiatives of this trend already appeared in Europe, in the birthplace of Enlightenment. Currently, there are already 25 European nations trying to harmonize their activities and organize a viable community. Nevertheless, the power of tradition, the heterogeneity of inherited Weltanschauungs, the usually misinterpreted national interests, and so on will for a good while successfully slow the final victory of this irresistible trend.

*

The purposeful manipulation of the human brain (domestication) is the *sine qua non* for the establishment and maintenance of a community. As a rule, education from birth until sexual maturity determines irreversibly the most important goals of the individual. Exceptions are few. Those of the exclusive society who command over others and obtain via proper education the power over their subjects have always developed proper ideologies to efficiently exercise their power. They manipulate their dependants even to sacrifice their lives when needed to reach the goals set by the authorities. The billions who remained during the history of mankind untouched by their wartime killings of the masses of their innocent peers and were ready to die in the name of "God", "fatherland" and so on illustrate the consequences of the practically unlimited capacity of the human brain to fix ICRs and acquire drives.

Considering the present situation, it seems obvious that all over the world myths determine the world outlook of the masses. This is rooted in tradition. It is undeniably true that the development of science/technology automatically shifts the organization of life in the direction of reason, but the influence of tradition is enormous. It is enough to think about Japan during World War II to realize that a society can exploit science and technology without changing its tradition of manipulating the masses in a way that keeps them unaccustomed from thinking and acting in terms of pragmatic needs. Societies that allow their educational methods to be more deeply pervaded by the spirit of Enlightenment enable their citizens to become more accustomed to thinking and acting in terms of pragmatic needs than their peers living in other communities where the techniques of brain manipulation have remained more archaic.

The human brain at birth is ready to cram facts and develop the ability of individual thinking. Everything depends on education, on the way the brain is manipulated by the home/school/society triad during the most sensitive, uphill period of life, from weaning until sexual maturity. In societies ruled by dictators or controlled by collective dictators (church, army), individual thinking is suppressed. When they have power, they manipulate the brains of their dependants accordingly and brutally suppress individuals who deviate from the line.

Nevertheless, even democratic systems manipulate their dependants in a traditional manner. It is part and parcel of the history of mankind that in order to properly control the masses the élite has always curbed access to information. Even if we are aware of the highly significant difference between a democracy and a tyranny, we have to admit that secrecy and misinformation have always been and still are essential for the maintenance of the desired forms of community life. A revolutionary change in the general access to information is obviously of crucial importance for the future. Proper methods always develop when the time is ripe for them. For example, today through the *internet* everybody can easily and without delay get in contact with everybody and everybody has unlimited, easy access to everything whatsoever that human cortical activity has produced in the past of abiding value. Thus the interpersonal communication technique necessary for the development of a reason-directed human society has already been brought into the world.

It is obvious that the manipulated brain of the masses, filled from birth with emotional misbeliefs, illusions, misinterpretations, myths, fantasies, motivates – given the right circumstances – irrational activities. In peaceful periods of the community rational brain activity is dominant and the irrational, emotional chaos is suppressed. There are specially organized occasions for the eruption of the volcano (concerts, sport-events, proper political party meetings, etc.). But in emotionally supersensitive, trying periods of the community (riots, revolutions, wars) the irrational brain becomes dominant. The immense literature that describes the history of mankind is, as a matter of fact, an exact description of the consequences of traditional manipulation of the human brain.

The history of the myths-directed human society is a mine of the irrational consequences of misbeliefs traditionally handed down from generation to generation. It is enough to read a few of the more recent overviews: *Massacre in History* (Levene and Roberts, 1999), *In God's Name: Genocide and Religion in the Twentieth Century* (Bartov, 2001), *Modern Hatreds: The Symbolic Politics of Ethnic War* (Kaufman, 2001), *The Medical Documentation of Torture* (Iacopino and Peel, 2002) to get acquainted with the worst manifestations of malevolently manipulated human brains.

Even in the dark history of mankind, the Holocaust – the extermination of millions within a few years with unprecedented success, due to a systematically planned and executed evil mass manipulation of a whole nation – was a unique event. This horrifying example testifies to the fact that the potential to misuse the physiological endowments of the human cortex is practically unlimited.

It is a rule that intolerable social/political circumstances make the brain of masses highly susceptible for proper manipulation. The situation in Germany

5.2 The Timeliness of the Conception of the Enlightenment: *Sapere Ande!*

after World War I was intolerable for the masses. The history of Germany between 1933 and 1945 provides evidence of how far the fixation of hundreds of thousands or millions of ICR chains can lead, prompted by the belief that Germans belong to a chosen race that has to fulfill its destiny and play a dominant role in the community of nations.

Hitler – a fanatic, evil genius obsessed with a pathologic and inhuman idea – exploited the teachings of the home/school/society triad, handed down in Germany from generation to generation. He convincingly proved the manipulability at the basis of human brain physiology; he proved that on the basis of a masterfully applied series of manipulative techniques, the morals of a whole nation, previously highly respected for its achievements in science and art, can degenerate within a few years. The properly prepared masses, imbued by hatred, served as reliable tools to reach the evil aims that he set.

Anti-Semitism was Hitler's key instrument in the manipulation of his subjects' brains. The fate of Jews in the Christian world, culminating in the Holocaust, is an especially tragic memento, demonstrating irrationalism in the myths-directed world. Neither the Holocaust nor the attitude of the European countries with regard to the current Arab-Israeli conflict is understandable without the sober analysis of the irrational fate of the Jewish community in the European countries after the rise of Christianity. To throw light on the completely irrational nature of this problem we need to sum up, as briefly as possible, the most important facts described in detail in the practically unlimited literature dealing with this subject.

Based on the misinterpretation of a historical event it became a basic educational method in Christian societies to fix from birth millions of chains of ICRs inseparable from the cognitive and affective states of consciousness described as anti-Semitism into the brain of their subjects. As a consequence of this indoctrination Jews were already expelled in 1290 from England, in 1394 from France, and in 1492 from Spain. For centuries they were discriminated against and cruelly treated in European countries.

But even the emancipation of the Jews, first in France in 1791, did not radically change the traditional spirit of education in the Christian world. Its consequences are, up to the present day, easily traceable. It is not only strangely unusual but even almost tragicomic that of the 175 UN Security Council resolutions passed before 1990, 97 were directed against Israel. Of the 690 General Assembly resolutions voted on before 1990, 429 were directed against Israel: *A country only slightly larger than the Canary Islands!* On the other hand, the UN was silent while in Jordan: (a) 58 Jerusalem synagogues were destroyed, (b) the ancient Jewish cemetery on the Mount of Olives was desecrated, and (c) Jews were prevented from visiting the Temple Mount and the Western Wall.

Israel became a nation in 1312 B.C.E. Since the Jewish conquest in 1272 B.C.E., the Jews had had dominion over the land for one thousand years with a continuous presence in the land for the past 3,300 years. The only Arab-dominion since the conquest in 635 C.E. lasted no more than 22 years. For over 3,300 years, Jerusalem has been the Jewish capital. The city is mentioned over 700 times in the Jewish Holy Scriptures, but is not mentioned once in the Koran. King David founded the city and Mohamed never came to Jerusalem. Jews pray facing Jerusalem. Muslims pray with their back toward Jerusalem. Under Jordanian rule, Jewish holy sites were desecrated and the Jews were denied access to places of worship. Under Israeli rule, all Moslem and Christian sites have been preserved and made accessible to people of all faiths. The Arabs are represented by eight separate nations, not including the Palestinians. There is only one Jewish nation. Arab nations have initiated five wars against Israel and lost. In 1948 the Arab refugees, estimated to be around 630,000, were encouraged to leave Israel by Arab leaders. Sixty-eight percent left without ever seeing an Israeli soldier. Jewish refugees, estimated to be around 600,000, were forced to flee from Arab lands, due to brutality, persecution, and pogrom. Out of the hundred million refugees since World War II, Arab refugees represent the only group in the world that has *intentionally* never been absorbed or integrated into their own people's land.

Living in ghettos and being compelled to adapt themselves to their highly specific conditions, the chains of ICRs and acquired drives from birth, transmitted for centuries from generation to generation is a unique cultural heritage of the Jewish people. These ICRs were significantly different from those fixed in their neighbors' brains. In retrospect we may say that to survive Jews were always required to better exploit the physiological endowments of their brains. Rich or poor, the leitmotif of education was always, from the time of the ghettos to the post-emancipatory period, "try to be much better than the average" and thus learn, learn, learn. In times when the overwhelming majority in the Christian communities was illiterate, most of the children in the Jewish communities learned reading and writing. It is no mere coincidence that the State of Israel leads the world in the number of scientists and technicians in the workforce, with 145 per 10,000, as opposed to 85 in the United States, over 70 in Japan, and less than 60 in Germany; it also has more museums per capita than any other country. The unusually high number of Nobel Prize winners of Jewish origin is also not accidental. Thus, from those Jewish scientists (and their children) who escaped to England, 16 received the Nobel Prize and 72 became members of the Royal Society.

The "Jewish problem", its terrible outgrowths in the 20th century, and the new rising phase of anti-Semitism in Europe 60 years after the Holocaust are all horrifying, yet I consider them as natural symptoms, concomitant signs, typical outward manifestations of a myths-directed society. Unfortunately, anti-Semitism is just one of the historically developed irrational hatreds with which the still imbalanced present world of the *Homo sapiens sapiens* is imbued.

5.2 The Timeliness of the Conception of the Enlightenment: *Sapere Ande!* 137

Only a global change of education based fully on the exact knowledge of the brain mechanisms that enable the manipulation of individuals can lead, at some point in the future, to the desired rationally directed society.

*

The careful analysis of the development of the glass-cylinder-seeking drive revealed that mammals capable of acquiring drives (domesticable animals and humans) are born with naive cortical neurons the functional significance of which is changed through a proper learning procedure. The trained neurons become involved either in an ECR, an ICR, or an acquired drive. It is the essence of the life of mammals capable of acquiring drives that a perpetually growing number of naive cortical neurons learn to work in concert with each other and change their functional significance from birth until death. It is obvious that neurons, designed for lifelong operation, are nonproliferating cells.

Based on experience, humans have domesticated animals from ancient times to exploit them thereafter, and human societies have domesticated humans from ancient times to exploit them thereafter. In advanced human societies with a high level of culture, governmental methods developed on the basis of influential mythical ideologies have, for thousands of years, served the aim: (a) of keeping the members of the community together and (b) of fostering cooperation to create the proper conditions for development. Thus, the manipulative techniques served not only the élites, but also the interest of the community.

This situation changed dramatically toward the end of the 20th century. Countries that developed rapidly owing to their democratic systems reached a high level in science and technology and improved significantly the quality of life of their citizens. Unfortunately, because of their old traditions, they also produced extremely dangerous weapons of mass destruction. Nevertheless, owing to the fine regulations of democratic systems, these weapons are reasonably controlled. It is almost impossible, however, to prevent communities lacking such controls and driven by ancient religious misbeliefs from getting access to such sophisticated weaponry and misusing it.

Though, to date, the misuse of nuclear weapons has fortunately been prevented, the special new, cheap but highly dangerous weapon, the properly manipulated suicide killer, now keeps the civilized world increasingly concerned and intimidated. Their terrifying methods, demonstrated with special brutality in New York on September 11, 2001, in Madrid on March 11, 2004, or in Beslan (Russia) on September 1, 2004, rightly frighten the developed societies all over the world. To reestablish peaceful order in the world, leading powers try to use traditional means of coercion to conquer the new enemy. The traditional methods of war are, however, obviously ineffective. The new situation in the world is the telltale sign that the first phase of history, the myths-directed human society is approaching its end. It is the order of our age that a rationally directed world must replace our present myths-directed one, to prevent forever the proliferation of maliciously manipulated human brains and their treacherous weapon, from which there is no escape.

A community whose members need to cooperate to reach their common goals has to be held together by a common or an essentially similar world outlook. Human society has been operating since tens of thousands of years. Thus, given the lack of exact knowledge about the rational work of the cortex, it was unavoidable that the world outlook was built upon myths, upon the chaos in our brain. I conjecture that this first period of the development of the human society will necessarily be terminated following the understanding of (a) the force of mutual attraction between cortical neurons and (b) the clarification of the chemical nature of cortical enhancer regulation, resulting in the fixation of ICRs and acquired drives. The final equilibrium of human society will rest on the firm knowledge that (a) human beings are theoretically all equal, since each of their brains possesses the endowments that make them potentially able to produce everything that humans have ever produced in the past and will ever produce in the future, (b) the immense capacity of the human brain is exploitable, (c) everybody needs to strive consciously after the acquisition of drives best fitting to his/her innate endowments, and (d) the community is obliged to give maximum support to reach affordable goals.

To replace the presently dominant myths-created heterogeneous forms of Weltanschauung with an homogeneous one based on brain research will obviously represent a radical turning point in the history of mankind. Its supporting pillar remains the rule that *human life and the dignity of the individual is inviolable.*

With the separation of Church and State, the Enlightenment served as catalyst, more than 200 years ago, for a step in the right direction. Every democratic country today grants their citizens legal protection to decide freely and without any brutal adverse consequences their relation to Church and God. Nevertheless, the spirit of education, imbued with traditional myths, did not change radically.

The missing condition for a generally acceptable rational world outlook is the lack of knowledge about the specific chemistry in the brain when an acquired-drive-directed goal-seeking behavior is in progress. Only the ultimate clarification of the force of attraction between cortical neurons can prevent, sometime in the future, the uncontrolled proliferation of fanatic brains and create a global human society whose members can consciously strive to acquire drives best fitting to their natural endowments. Taught correctly in due time, people will have the chance to select – from the immense reservoir of beauty created by art – the most enjoyable for them, and likewise to understand those achievements of science that find an easy way into their mind. Therefore whoever has the inner drive will be able to add his or her own contribution to the already available values. Paraphrasing Goethe's wise remark, "He who possesses art or science has his religion"; he who builds the proper acquired

5.2 The Timeliness of the Conception of the Enlightenment: *Sapere Ande!*

drives into his brain has his religion. Most importantly, people with drives best fitting to their natural endowments have the best chance to be confident and happy, and to be free from the need for the so-called ersatz.

A glass-cylinder-seeking rat will never acquire this drive under natural conditions. We manipulated its brain, making use of the potential to change a group of cortical neurons via proper training. The best-performing glass-cylinder-seeking rat, which was initially trained under pressure and always acted under coercion in the learning phase, ultimately appears to have a fanatical desire to search the glass cylinder and jump onto its top rim. In 2% of the rats the circumstances in the brain were so favorable for the acquisition of the glass-cylinder-seeking drive that the latter even suppressed innate drives, and these animals go through "fire and water" to reach their unnatural goal.

Humans possess the most manipulable brain among all living creatures on earth. The brain of a suicide killer is furtively manipulated. The properly acquired drive develops as a result of long-lasting training. The subject always acts under coercion, under severe mental pressure. Nevertheless, *it is the nature of an acquired drive that, if the manipulation was fully successful, the individual ultimately behaves as one possessing a fanatical desire to reach the acquired-drive-motivated goal.* Just as our glass-cylinder-seeking rat behaves as one who has a fanatical desire for the glass cylinder.

An acquired drive is always built on an inner drive but after it has been ultimately fixed, the innate drive cannot be recognized any more. It is impossible to recognize either in man or in a domesticated animal the origin of an acquired drive.

Psychologists describe the behavior of human beings so vividly today that the reader develops a satisfying feeling of fully understanding what happens. As a matter of fact, the analysis of the behavior of a glass-cylinder-seeking rat invites a new interpretation. We assume that the training-induced transition of a naive cortical neuron until it belongs to Group 3 or 4 is the same in domesticable animals and humans, and the activity of neurons belonging to Group 3 or 4 is inseparable from conscious perception. It is just one example of the passing of quantity into quality that makes the performance of the human brain unique.

Conventional human thinking today is in harmony with *The Self and Its Brain* conception as described so elegantly and delightfully as interaction between Worlds 1, 2, and 3 (Popper and Eccles, 1977). In the light of the operation of mesencephalic and cortical enhancer regulation, the opposite conception, with *The Brain and Its Self* approach, deserves serious consideration.

*

Humans to survive needed reason, worked hard, and developed science; to relax, they developed art. For a full and happy life humans really need science to cope with harsh reality and also need art to escape from it. But they cannot be mixed. Myths, supernatural forces, and so on, even if they are described in religious holy scriptures, belong to art. A new global education must be based on science; on the knowledge of the crucial role of the ability to fix ICRs and acquire drives in the evolution of community life; on the knowledge of the physiological mechanisms which make the misuse of the human brain possible; and on the knowledge of how the immense capacity of our brain can be properly used.

It is obvious that the effort to eliminate the traditional mythical approach to human life will meet fierce opposition even in the 21st century. This goal, however, is unavoidable if one is to reach, in the long run, the revolutionary aim set more then 200 years ago by the brilliant pioneers of the Enlightenment. Their approach, *Sapere ande!* (Dare to go independently), is as timely as it was then. If masses learn how their brain works, they will resist traditional methods of manipulation.

*

Gauguin's monumental painting (now in the Museum of Fine Arts, Boston) raises the three *fundamental* questions of humans that we may answer, in the light of the approach presented in this study, as follows:

1. *Where Do We Come From?*
   The chance evolution of chemical reactions led to the birth of "living material", complex molecules capable of self-reproduction. With the passing of time , the natural trial-and-error process of development necessarily produced more and more sophisticated organisms, and this manifested itself by growth through metabolism, reproduction, and the power of adaptation to the environment through changes originating internally. Ultimately species whose individuals were capable of cooperating with each other appeared, and this new line of development arrived at its present peak with the myths-directed human society.

2. *What Are We?*
   The *Homo sapiens sapiens* is a unique species whose activity is primarily determined by acquired drives and has therefore reached the most sophisticated forms of social life found to date.

3. *Where Are We Going?*
   Since the revolution of the Enlightenment that opened the way for rapid progress in science/technology, the myths-directed, chaotically heterogeneous communities of humans are still in the stage of transformation directed towards the irreversible establishment of a rationally directed, homogeneous human society.

6 Conclusion

The main message of this monograph is that the appearance of the mammalian brain with the ability to acquire drives ensured the development of social life and ultimately led to the evolution of human society. This most sophisticated form of organized life on earth is still in the trial-and-error phase of its development. It seeks to outgrow the myths-directed era of its history and arrive at its final state, the rationally directed human society.

To study the formation of an acquired drive, a rat model was developed. The "glass-cylinder-seeking drive" was fixed into the brain of rats. Based on an unconditioned avoidance reflex (escape from a hot plate) and using the sound of a shrill bell, playing the role of a high-priority conditioned stimulus, rats were trained to search for and jump to the rim of a 30-cm-high glass cylinder open at bottom and top with diameters of 16 cm and 12 cm, respectively, and with a side opening through which a rat (up to 350–400 g body weight) manages to get inside the cylinder. The best-performing rats acquired the glass-cylinder-seeking drive in a stable manner, possessing thereafter this unnatural urge for a lifetime. These rats showed the same high-grade adaptability and readiness in overcoming different obstacles while reaching the goal as the ones influenced by innate drives, such as hunger or sexual desire.

It was shown that the faculty for acquiring a drive is uncommon in the animal kingdom. Vertebrates can be divided into three groups according to the mode of operation of their brain: (a) those which operate with innate drives only (the majority), (b) those with an ability to acquire drives (a minority), and (c) the only one which operates almost exclusively on the basis of acquired drives (*Homo sapiens*).

With the evolution of brains capable of acquiring drives, species appeared whose members could manipulate each other's behavior and act together. This was the condition *sine qua non* for the evolution of social living, a form of life that enabled the social group to surpass qualitatively the performance of the individual. It goes without saying that the training of skills to act in concert improved the quality of life.

With the development of the human brain, a functional network with over 100 billion interrelated nerve cells and 10^{10} bit capacity arose. With this system, from the operation of which conscious perception is inseparable, life on earth reached its most sophisticated form of appearance. Furthermore, the human being, primarily a social creature, is a building block in the creation of a gigantic product: human society. The function and capacity of society obviously exceeds the sum of the activity of its members. Based on the practically inexhaustible capacity of the human brain to acquire drives, human society represents a qualitatively new, higher form of life.

Decades of studies into the nature of the development of an acquired drive revealed the following:

In the mammalian brain capable of acquiring drives, untrained cortical neurons (Group 1) possess the potentiality to change their functional state in response to practice, training, or experience in three consecutive stages, namely, by getting involved in (a) an extinguishable conditioned reflex (ECR) (Group 2), (b) an inextinguishable conditioned reflex (ICR) (Group 3), or (c) an acquired drive (Group 4). The activity of the cortical neurons belonging to Groups 3 and 4 is inseparable from conscious perception. In any moment of life self is the sum of those cortical neurons that have already changed their functional significance and belong to Groups 3 or 4.

Thinking over the timetable of the development of life on earth we realize suddenly the unique significance of the appearance of the human brain. Unicellular organisms are 5–1.5 billion years old, modern mammals appeared 60–40 million years ago, and *human society is only about 80,000 years old*. It is still in the trial-and-error phase of its development seeking its final equilibrium state.

Metaphorically, every human being is born with a telencephalon that resembles a book with over 100 billion empty pages (untrained, naive cortical neurons, Group 1), and with the capacity to inscribe as much as possible in this book throughout life. In reality, cortical enhancer regulation – the modification of the presently still unknown chemistry of the cortical neurons through learning, aiming to establish cooperation between cortical neurons previously unacquainted with one another – is the essence of human life. *Whenever a drive is acquired, chains of ICRs are fixed, neurons responsible for emotions are also coupled to the integral whole, thus cognitive/volitional consciousness is necessarily inseparable from an affective state of consciousness.* The mechanism that binds emotions as appurtenances to any chain of ICRs is of crucial importance to interpersonal communication.

Cortical neurons belonging to Group 3 or 4 continuously synthesize their specific enhancer substance within their capacity. This means that even in the vigilant resting state (leisure), in the absence of a dominant drive, as well as in the nonvigilant resting state (sleeping), the cortical neurons representing the totality of the already fixed ICRs and acquired drives are permanently under the influence of their specific enhancer substance. Although the level of this permanent, undulating activation remains low, it is unpredictable as to when any group of cortical neurons will be influenced by enhancer substances on the level already inseparable from conscious perception. Thus, as the totality of the cortical neurons belonging to Group 3 or 4 works continuously on an

6 Conclusion

unconscious level, there is a steadily operating, chaotic background noise in the human telencephalon. Although in the active state ("fight or flight" behavior, goal-seeking), when the actually dominant drive determines the rational goal to be reached, and the background noise is suppressed, it can never cease to exist. But it never endangers the function of the actually dominant innate or acquired drive. From this situation it follows that *rational brain activity is necessarily amalgamated with irrational brain activity and we live through every moment of our life experiencing the totality of order and chaos in our brain.*

Thus, (a) whenever a chain of ICRs is fixed in the human brain, the proper cortical neurons remain, on an unconscious level, constantly active for life, and (b) if the proper method is used, even a chain of ICRs that had never been ecphorized after fixation can be activated to the level needed for conscious perception at any later date. The recalling of any chain of ICRs is necessarily inseparable from an affective state of consciousness, due to the emotions coupled as appurtenances to the cortical neurons when they learned to cooperate with each other. Freud developed empirically sound methods for ecphorizing forgotten chains of ICRs in humans, decades after their fixation.

All in all, (a) past experiences are irreversibly fixed in neurons belonging to Groups 3 and 4 that learned to cooperate with each other and constitute an integral whole and (b) proper stimulation of the cooperating neurons as an integral whole allows the fixed information to be ecphorized any time later. This is inseparable from conscious perception, and thus the past experience is vividly re-lived in a cognitive and affective manner. (c) Even though it is true that during the operation of a dominant drive the activity of the individual is primarily focused on reaching the goal represented by this drive (rational activity), the ability to simultaneously consciously revive past experiences that are outside the limits of the dominant drive actually operating (irrational activity) is a natural endowment of the brain.

Because of the theoretically immense variability of cortical enhancer regulation, any trial to develop a compound that will reasonably stimulate learning in general seems to be, from physiological point of view, a hopeless undertaking. The natural method of behavioral modification – by means of experience, training, or practice – will likely remain not only the most effective, but the only viable way to change the performance of the cortical neurons in species capable of acquiring drives, and this presumably forever,. Everything depends therefore, and will probably always depend, on teaching, learning, education.

As the background noise in the brain is never interrupted and can even become more accentuated during sleep than in the vigilant resting state, the dreamworld, the classic example of a man-created universe, has always given inspiration to art. Its ultimate explanation awaits the resolution of scientific problems: the chemistry of cortical enhancer regulation and the natural law that determines the operation of the brain and its self and is responsible for the immense variability of human activities.

Human society – the maintenance of which has always required the proper manipulation of the brain of its members – still finds itself in a state of development. It seeks its final equilibrium: namely, that state in which behavioral modification induced by the home/school/society triad will be based, from birth until death, on the exact knowledge of the natural laws that keep the brain and its self going. In this way, members of the community will understand that simultaneity of order and chaos in their brain is the physiological reality that determines human activity, and will consciously try to find the acquired drives that optimally fit their natural endowments.

For the time being those who have been lucky enough to acquire the best fitting drives in due time, in the early uphill period of life, have had fair chances for success and happiness. In contrast, those who for any reason have missed this opportunity will remain frustrated and look for "ersatz". It seems reasonable to conclude that order and chaos are of equal importance in our brain. *Without the ability to adapt ourselves to the concrete (science), we would not be able to survive; without the ability which allows detachment from the concrete and explorations in the infinite (art), life would not be worth living.*

The existence of a variety of animal species with extremely restricted abilities to fix ICRs and acquire drives marks Nature's long road of experimentation with the brain. The end result of this process has been the most perfect variety. About 40,000 years ago, the *Homo sapiens sapiens* appeared and the progressive development of human society, capable of consciously changing its essential conditions, began. The limitless capacity of the human cortex to fix ICRs and acquire drives allowed, in conjunction with the development of language, an unmatched interpersonal communication. This unique facility made the cognitive/volitional and affective states of consciousness of the human brain, and as a consequence of it, human social life, unparalleled. Because of the lack of similar developments in animals, there is no way to get direct evidence of the nature of their psychic experiences. Nevertheless, observation of the goal-seeking behavior of properly trained monkeys, dogs, horses, and so on furnishes convincing indirect evidence for the assumption that the operation of ICRs and acquired drives is inseparable, even in animals, from an archetype of consciousness.

Since its birth, the myths-directed human community has undergone progressive transformation. It still seeks its final state of equilibrium: a society whose organization is fully rational, based on knowledge of the natural laws that keep the brain and its self operating. Regarding the neurochemical basis of the drives, experimental evidence and theoretical considerations in this monograph have led us to conceptualize that an until recently unknown brain mechanism, enhancer regulation in the brain stem, is primarily responsible for the innate drives, and a special form of it in the cortex is primarily responsible for the acquired drives.

6 Conclusion

Enhancer regulation is defined as: the existence of enhancer-sensitive neurons capable of changing their excitability and working on a higher activity level in a split second, due to endogenous enhancer substances. Of these substances, β-phenylethylamine (PEA) and tryptamine are currently being experimentally analyzed, and their synthetic analogues, (−)-deprenyl and (−)-BPAP, respectively, are the specific experimental tools for studying enhancer regulation in the brain stem.

Furthermore, data support the conclusion that age-related changes in enhancer regulation of the catecholaminergic brain engine are primarily responsible for: (a) the youthful power of mammals from weaning until sexual maturity, (b) the transition from the uphill period of life into postdevelopmental longevity, (c) the progressive decay of behavioral performances during the downhill period, and (d) the transition from life to death. Experimental and clinical studies with (−)-deprenyl, at present the only synthetic mesencephalic enhancer substance in clinical use, strongly support the proposal that prophylactic administration of a synthetic mesencephalic enhancer substance during postdevelopmental life could significantly slow the unavoidable decay of behavioral performances, prolong life, and prevent or delay the onset of age-related neurodegenerative diseases such as Parkinson's and Alzheimer's.

References

Ajika K, Hökfelt T (1973) Ultrastructural identification of catecholamine neurons in the hypothalamic preventricular arcuate nucleus-median eminence complex with special reference to quantitative aspects. Brain Res 57: 97–117

Alajouanine T (1948) Aphasia and artistic realization. Brain 71:229–241

Albin RL, Young AB, Penney JB (1989) The functional anatomy of basal ganglia disorders. Trends Neurosci 12:366–375

Allain H, Gougnard J, Naukirek HC (1991) Selegiline in de novo parkinsonian patients: the French selegiline multicenter trial (FSMP). Acta Neurol Scand 136:73–78

Amsterdam JD (2003) A double-blind, placebo-controlled trial of the safety and efficacy of selegiline transdermal system without dietary restrictions in patients with major depressive disorder. J Clin Psychiatry 64:208–214

Ban TA, Healy D, Shorter E (1998) The rise of psychopharmacology and the story of CINP. Animula Publishing House, Budapest

Bartov O (2001) In God's name: genocide and religion in the twentieth century (War and genocide, vol 4). Berghahn Books

Beritov IC (1932) As cited in Knoll J (1969) The theory of active reflexes. Hungarian Academy of Sciences, Budapest; Hafner Publishing, New York, p 10

Bertler A (1961) Occurrence and localization of catecholamines in human brain. Acta Physiol Scand 51:135–161

Birkmayer W, Hornykiewicz O (1962) Der L-Dioxyphenyl-Alanin-Effekt beim Parkinson-Syndrom des Menschen. Arch Psychiat Nervenkrh 203:560–564

Birkmayer W, Riederer P, Ambrozi L, Youdim MBH (1977) Implications of combined treatment with "Madopar" and L-deprenil in Parkinson's disease. Lancet 1:439–443

Birkmayer W, Riederer P, Linauer W, Knoll J (1984) L-Deprenyl plus L-phenylalanine in the treatment of depression. J Neural Transm 59:81–87

Birkmayer W, Knoll J, Riederer P, Youdim MBH, Hars V, Marton V (1985) Increased life expectancy resulting from addition of L-deprenyl to Madopar treatment in Parkinson's disease: a long-term study. J Neural Transm 64:113–127

Birks J, Flicker L (2003) Selegiline for Alzheimer's disease. Cochrane Database Syst Rev 1:CE000442

Blackwell B (1963) Hypertensive crisis due to monoamine oxidase inhibitors. Lancet ii:849–851

Bodkin JA, Amsterdam JD (2002) Transdermal selegiline in major depression: a double-blind, placebo-controlled, parallel-group study in outpatients. Am J Psychiatry 159:1869–1875

Bookheimer SY, Strojwas MH, Cohen MS, Saunders AM, Pericak-Vance MA, Mazziotta JC, Small GW (2000) Patterns of brain activation in people at risk for Alzheimer's disease. N Engl J Med 343:450–456

Borowsky B, Adham N, Jones KA, Raddatz R, Artimishyn R, Ogozalek KL, Durkin MM, Lakhlani PP, Bonini JA, Pathirana S, Boyle N, Pu X, Kouranova E, Lichtblau H, Ochoa FY, Branchek TA, Gerald C (2001) Trace amines: Identification of a family of mammalian G protein-coupled receptors. Proc Nat Acad Sci USA 98: 8966–8971

Borroni B, Archetti S, Agosti C, Akkawi N, Brambilla C, Caimi L, Caltagirone C, Di Luca M, Padovani A (2004) Intronic CYP46 polymorphism along with ApoE genotype in sporadic Alzheimer Disease: from risk factors to disease modulators. Neurobiol Aging 25:747–751

Boulton AA, Juorio AV, Downer RGH (1988) Trace amines: comparative and clinical neurobiology (Experimental and clinical neuroscience). Humana Press, Totowa, NJ

Bunzow JR, Sonders MS, Arttamangkul S, Harrison LM, Zhang G, Quigley DI, Darland T, Suchland KL, Pasumamula S, Kennedy JL, Olson SB, Magenis RE, Amara SG, Grandy DK (2001) Amphetamine, 3,4-methylenedioxymethamphetamine, lysergic acid diethylamide, and metabolites of the catecholamine neurotransmitters are agonists of a rat trace amine receptor. Mol Pharmacol 60:1181–1188

Calas A, Alonso G, Arnauld E, Vincent JD (1974) Demonstration of indolaminergic fibers in the median eminence of the duck, rat and monkey. Nature 250:242–243

Calas A, Besson NJ, Cauchy C, Alonso G, Glowinsky J, Cheramy A (1976) Radioautographic study of in vivo incorporation of ^{3}H-monoamines in the cat caudate nucleus:identification of serotoninergic fibers. Brain Res 118:1–13

Campi N, Todeschini GP, Scarzella L (1990) Selegiline versus L-acetylcarnitine in the treatment of Alzheimer-type dementia. Clin Ther 12:306–314

Campion D, Dumanchin C, Hannequin D, Dubois B, Belliard S, Puel M et al. (1999) Early-onset autosomal dominant Alzheimer disease: prevalence, genetic heterogeneity, and mutation spectrum. Am J Hum Genet 65:664–670

Carlsson A (1979) The impact of catecholamine research on medical science and practice. In: Usdin E, Kopin IJ, Barchas J. (eds) Catecholamines: basic and clinical frontiers, vol 1. Pergamon Press, New York, pp 4–19

Carrillo MC, Kanai S, Nokubo M, Kitani K (1991) (−)Deprenyl induces activities of both superoxide dismutase and catalase but not of glutathion peroxidase in the striatum of young male rats. Life Sci 48:517–521

Carrillo MC, Kanai S, Nokubo M, Ivy GO, Sato Y, Kitani K (1992) (−)Deprenyl increases activities of superoxide dismutase and catalase in striatum but not in hippocampus: the sex and age-related differences in the optimal dose in the rat. Exp Neurol 116:286–294

Chan-Palay V (1975) Fine structure of labelled axons in the cerebellar cortex and nuclei of rodents and primates after intraventicular infusions with tritiated serotonin. Anat Embryol 148:235–265

Cohen GP, Pasik B, Cohen A, Leist C, Mitileneou MD, Yahr M (1984) Pargyline and (−)deprenyl prevent the neurotoxicity of 1-methyl-4-phenyl-1,2,3,6-tetrahydropyridine (MPTP) in monkeys. Eur J Pharmacol 106:209–210

Connick JH, Hanlon G, Roberts J, France L, Fox PK, Nicholson CD (1992) Multiple sigma binding sites in guinea-pig and rat brain membranes: G-protein interactions. Br J Pharm 107:726–731

Crane GE (1956) Psychiatric side effects of iproniazid. Am J Psychiatr 112: 494–499

Dabrowsky M (1995) Kandinsky compositions. The Museum of Modern Art, New York

Dalló J, Lekka N, Knoll J (1986a) Age dependent decrease of copulatory activity and its correction by (−)deprenyl in male rats. In: Borsy J, Kerecsen L, György L (eds) Dopamine, ageing and diseases. Pergamon Press, Akadémiai Kiadó, Budapest, pp 35–38

Dalló J, Lekka N, Knoll J (1986b) The ejaculatory behavior of sexually sluggish male rats treated with (−)deprenyl, apomorphine, bromocriptine and amphetemine. Pol J Pharmacol Pharm 38: 251–255

Davis BA, Boulton AA (1994) The trace amines and their acidic metabolites in depression an overview. Prog Neuro-Psychopharmacol Biol Psychiatry 18:17–45

Descarries L, Léger L (1978) Serotonin nerve terminals in the locus coeruleus of the adult rat. 3. In: Garattini S, Pujol JF, Samanin R (eds) Interactions between putative neurotransmitters in the brain. Raven Press, New York, pp 355–367

Descarries L, Watkins KC, Lapierre Y (1977) Noradrenergic axon terminals in the cerebral cortex of rat. Topmetric ultrastructural analysis. Brain Res 133:197–222

Dobbs SM, Dobbs RJ, Charlett A (1996) Multi-centre trials: U-turns by bandwagons and the patient left by the wayside. Br J Clin Pharmacol 42:143–145

Eckert B, Gottfries CG, Knorring L, Oreland L, Wilbert A, Winblad B (1980) Brain and platelet monoamine oxidase in schizoprenics and cycloid psychotics. Prog Neuropsychopharmacol 4:57–68

Elsworth JD, Glover V, Reynolds GP, Sandler M, Lees AJ, Phuapradit P, Shaw KM, Stern GM, Kumar P (1978) Deprenyl administration in man; a selective monoamine oxidase B inhibitor without the "cheese effect". Psychopharmacology 57:33–38

Etminan M, Gill S, Samii A (2003) Effect of non-steroidal anti-inflammatory drugs on risk of Alzheimer's disease: systematic review and meta-analysis of observational studies. Brit Med J 327:128–131

Falsaperla A, Monici Preti PA, Oliani C (1990) Selegiline versus oxiracetam in patients with Alzheimer-type dementia. Clin Ther 12:376–384

Finnegan KT, Skratt JJ, Irvin I, DeLanney LE, Langston JW (1990) Protection against DSP-4 induced neurotoxicity by deprenyl is not related to its inhibition of MAO-B. Eur J Pharmacol 184:119–126

Fischer E, Heller B, Miró AH (1968) β-Phenylethylamine in human urine. Arzneimittelf 18:1486

Fischer E, Spatz H, Heller B, Reggiani H (1972) Phenethylamine content of human urine and rat brain, its alterations in pathological conditions and after drug administration. Experientia 15:307–308

Fowler CJ, Oreland L, Marcusson J, Winblad B (1980) Titration of human brain monamine oxidase -A and -B by clorgyline and L-deprenyl. Naun-Schmied Arch Pharmacol 311:263–272

Fowler CJ, Wiberg A, Oreland L, Marcusson J, Windlab B (1980) The effect of age on the activity and molecular properties of human brain monoamine oxidase. J Neural Transm 49:1–20

Greenshow AJ (1989) Functional interactions of 2-phenylethylamine and of tryptamine with brain catecholamines: implications for psychotherapeutic drug action. Prog Neuro-Psychopharmacol Biol Psychiatry 13:431–443

Grundman M (2000) Vitamin E and Alzheimer disease: the basis for additional clinical trials. Am J Clin Nutr 71:630s–636s

Hamabe W, Fujita R, Yasusa T, Yoneda F, Yoshida A, Ueda H (2000) (−)1-(Benzofuran-2-yl)-2-propylaminopentane shows survival effect on cortical neurons under serum-free condition through sigma receptors. Cell Mol Neurobiol 20:695–702

Hársing RG, Magyar K, Tekes K, Vizi ES, Knoll J (1979) Inhibition by (−)-deprenyl of dopamine uptake in rat striatum: A possible correlation between dopamine uptake and acetylcholine release inhibition. Pol J Pharmacol Pharm 31:297–307

Hauger RL, Skolnick P, Paul SM (1982) Specific [^{3}H] beta-phenylethylamine binding sites in rat brain. Eur J Pharmacol 83:147–148

Healy D (1996) The psychopharmacologists I. Altman, London, Weinham, New York, Tokyo, Melbourne, Madras

Healy D (1998) The psychopharmacologists II. Altman, London, Weinham, New York, Tokyo, Melbourne, Madras

Healy D (2000) The psychopharmacologists III. Arnold. London, Oxford University Press Inc., New York

Helmer C, Joly P, Letenneur L, Commenges D, Dartigues JF (2001) Mortality with dementia: results from a French prospective community-based cohort. Am J Epidemiol 154:642–648

Hilgard ER (1948) Theories of learning. Appleton Century Crofts, New York

Hökfelt T (1968) In vitro studies on central and peripheral monoamine neurons at the ultrastructural level. Z Zellforsch 91:1–74

Hy LX, Keller DM (2000) Prevalence of AD among whites: a summary by levels of severity. Neurology 55:198–204

Iacopino V, Peel M (2002) The medical documentation of torture. Greenich Medical Media, London

Janssen PA, Leysen JE, Megens AA, Awouters HF (1999) Does phenylethylamine act as an endogenous amphetamine in some patients? Int J Neuropsychopharmacol 2:229–240

Johnston JP (1968) Some observations upon a new inhibitor of monoamine oxidase in human brain. Biochem Pharmacol 17:1285–1297

Kaufman S (2001) Modern hatreds: The symbolic politics of ethnic war. Cornell Univ Press

Kelemen K, Longo VG, Knoll J, Bovet D (1961) The EEG arousal reaction in rats with extinguishable and non-extinguishable conditioned reflexes. Electroenc Clin Neurophysiol 13:745–751

Kitani K, Kanai S, Sato Y, Ohta M, Ivy GO, Carrillo MC (1992) Chronic treatment of (−)deprenyl prolongs the life span of male Fischer 344 rats. Further evidence. Life Sci 52:281–288

Kitani K, Minami C, Isobe K, Maehara K, Kanai S, Ivy GO, Carrillo MC (2002) Why (−)deprenyl prolongs survival of experimental animals: Increase of anti-oxidant enzymes in brain and other body tissues as well as mobilization of various humoral factors may lead to systemic anti-aging effects. Mech Aging Dev 123:1087–1100

Kline NS (1958) Clinical experience with iproniazid (marsilid). J Clin Exp Psychophathol 19:72–81

Knoll B (1961) Certain aspects of the formation of temporary connections in comparative experiments on mice and rats. Acta Physiol Hung 20:265–275

Knoll B (1968) Comparative physiological and pharmacological analysis of the higher nervous function of mice and rats (in Hungarian). Candidate of Sciences Thesis (Ph.D equivalent). Hungarian Academy of Sciences, Budapest

Knoll J (1956) Experimental studies on the higher nervous activity of animals. V. The functional mechanism of the active conditioned reflex. Acta Physiol Hung 10:89–100

Knoll J (1957) Experimental studies on the higher nervous activity of animals. VI. Further studies on active reflexes. Acta Physiol Hung 12:65–92

Knoll J (1969) The theory of active reflexes. An analysis of some fundamental mechanisms of higher nervous activity. Hungarian Academy of Sciences, Budapest; Hafner Publishing Company, New York

Knoll J (1976) Analysis of the pharmacological effects of selective monoamine oxidase inhibitors. In: Wolstenholme GES, Knight J (eds) Monoamine oxidase and its inhibition. Ciba foundation Symposium 39 (new series), Elsevier, Amsterdam, pp 131–161

Knoll J (1978) The possible mechanism of action of (–)deprenyl in Parkinson's disease. J Neural Transm 43:177–198

Knoll J (1981a) Can the suicide inactivation of MAO by deprenyl explain its pharmacological effects? In: Singer TP, Ondarza N (eds) Molecular basis of drug action. Elsevier, Amsterdam, pp 185–201

Knoll J (1981b) The pharmacology of selective MAO inhibitors. In: Youdim, MBH, Paykel, ES (eds) Monoamine oxidase inhibitors: the state of the art. John Wiley and Sons, London, pp 45–61

Knoll J (1981c) Further experimental support to the concept that (–)deprenyl facilitates dopaminergic neurotransmission in the brain. In: Kamijo K, Usdin E, Nagatsu T (eds) Monoamine oxidase. basic and clinical frontiers. Excerpta Medica, Amsterdam, pp 230–240

Knoll J (1982) Selective inhibition of B type monoamine oxidase in the brain: a drug strategy to improve the quality of life in senescence. In: Keverling Buisman JA (ed) Strategy in drug research. Elsevier, Amsterdam, pp 107–135

Knoll J (1983) Deprenyl (selegiline). The history of its development and pharmacological action. Acta Neurol Scand Suppl 95:57–80

Knoll J (1985) The facilitation of dopaminergic activity in the aged brain by (–)deprenyl. A proposal for a strategy to improve the quality of life in senescence. Mech Ageing Dev 30:109–122

Knoll J (1986a) Striatal dopamine, aging and deprenyl. In: Borsy J, Kerecsen L, György L (eds) Dopamine, ageing and diseases. Pergamon Press, Akadémiai Kiadó (Hungarian Academy of Sciences), Budapest, pp 7–26

Knoll J (1986b) The pharmacology of (–)deprenyl. J Neural Transm Suppl 22:75–89

Knoll J (1986c) Role of B-type monoamine oxidase inhibition in the treatment of Parkinson's disease. An update. In: Shah NS, Donald AG (eds) Movement disorders. Plenum Press, New York, pp 53–81

Knoll J (1987) R-(–)Deprenyl (Selegiline, Movergan®) facilitates the activity of the nigrostriatal dopaminergic neuron. J Neural Transm 25:45–66

Knoll J (1988) The striatal dopamine dependency of lifespan in male rats. Longevity study with (–)deprenyl. Mech Ageing Dev 46:237–262

Knoll J (1989) The pharmacology of selegiline /(–)deprenyl. Acta Neurol Scand 126:83–91

Knoll J (1990) Nigrostriatal dopaminergic activity, deprenyl treatment, and longevity. Adv Neurol 53:425–429

Knoll J (1992a) Pharmacological basis of the therapeutic effect of (–)deprenyl in age-related neurological diseases. Med Res Rev 12:505–524

Knoll J (1992b) (–)Deprenyl-medication: a strategy to modulate the age-related decline of the striatal dopaminergic system. JAGS 40:839–847

Knoll J (1993a) The pharmacological basis of the beneficial effect of (–)deprenyl (selegiline) in Parkinson's and Alzheimer's diseases. J Neural Transm Suppl 40:69–91

Knoll J (1993b) The pharmacological basis of the therapeutic of (–)-deprenyl in age-related neurological diseases. In: Szelenyi I (ed) Inhibitors of monoamine oxidase B. Pharmacology and clinical use in neurodegenerative disorders. Birhkäuser Verlag, Basel, pp 145–168

Knoll J (1993c) Some clinical implication of MAO-B inhibition. In: Yasuhara H, Parvez SH, Oguchi K, Sandler M, Nagatsu T (eds) Monoamine oxidase: basic and clinical aspects. VSP Utrecht, The Netherlands, pp 197–217

Knoll J (1994) Memories of my 45 years in research. Pharmacol Toxicol 75:65–72

Knoll J (1995) Rationale for (–)deprenyl (selegiline) medication in Parkinson's disease and in prevention of age-related nigral changes. Biomed Pharmacother 49:187–195

Knoll J (1996) (–)Deprenyl (selegiline) in Parkinson's disease: a pharmacologist's comment. Biomed Pharmacother 50:315–317

Knoll J (1998) (–)Deprenyl (selegiline) a catecholaminergic activity enhancer (CAE) substance acting in the brain. Pharmacol Toxicol 82:57–66

Knoll J (2001) Antiaging compounds: (–)Deprenyl (Selegiline) and (–)1-(benzofuran-2-yl)-2-propylaminopentane, (–)BPAP, a selective highly potent enhancer of the impulse propagation mediated release of catecholamines and serotonin in the brain. CNS Drug Rev 7:317–345

Knoll J (2003) Enhancer regulation/endogenous and synthetic enhancer compounds: A neurochemical concept of the innate and acquired drives. Neurochem Res 28:1187–1209

Knoll J, Magyar K (1972) Some puzzling effects of monoamine oxidase inhibitors. Adv. Bioch Psychopharmacol 5:393–408

Knoll J, Miklya I (1994) Multiple, small dose administration of (–)deprenyl enhances catecholaminergic activity and diminishes serotoninergic activity in the brain and these effects are unrelated to MAO-B inhibition. Arch Int Pharmacodyn Thér 328:1–15

Knoll J, Miklya I (1995) Enhanced catecholaminergic and serotoninergic activity in rat brain from weaning to sexual maturity. Rationale for prophylactic (–)deprenyl (selegiline) medication. Life Sci 56:611–620

Knoll J, Kelemen K, Knoll B (1955a) Experimental studies on the higher nervous activity of animals. 1. A method for the elaboration of a non-extinguishable conditioned reflex in the rat. Acta Physiol Hung 8:327–345

Knoll J, Kelemen K, Knoll B (1955b) Experimental studies on the higher nervous activity of animals. 2. Differences in the state of function of the cells constituting the cortical representation of the unconditioned reflex in extinguishable and non-extinguishable conditioned reflexes. Acta Physiol Hung 8:347–367

Knoll J, Kelemen K, Knoll B (1955c) Experimental studies on the higher nervous activity of animals. 3. Experimental studies on the active conditioned reflex. Acta Physiol Hung 8:369–388

Knoll J, Kelemen K, Knoll B (1956) Experimental studies on the higher nervous activity of animals. 4. A method for elaborating and studying an active conditioned feeding reflex. Experimental analysis of differences between active conditioned defensive and feeding reflexes. Acta Physiol Hung 9:99–109

Knoll J, Ecseri Z, Kelemen K, Nievel J, Knoll B (1965) Phenylisopropylmethylpropinylamine (E-250) a new psychic energizer. Arch Int Pharmacodyn Thér.155:154–164

Knoll J, Vizi ES, Somogyi G (1968) Phenylisopropylmethylpropinylamine (E-250), a monoamine oxidase inhibitor antagonizing the effects of tyramine. Arzneimittelf 18:109–112

Knoll J, Yen TT, Dalló J (1983) Long-lasting, true aphrodisiac effect of (−)deprenyl in sexually sluggish old male rats. Mod Probl Pharmacopsychiatry 19:135–153

Knoll J, Dalló J, Yen TT (1989) Striatal dopamine, sexual activity and lifespan. Longevity of rats treated with (−)deprenyl. Life Sci 45:525–531

Knoll J, Knoll B, Török Z, Timár J, Yasar S (1992a) The pharmacology of 1-phenyl-2-propylaminopentane (PPAP), a deprenyl-derived new spectrum psychostimulant. Arch int Pharmacodyn Thér 316:5–29

Knoll J, Tóth V, Kummert M, Sugár J (1992b) (−)Deprenyl and (−)parafluorodeprenyl-treatment prevents age-related pigment changes in the substantia nigra. A TV-image analysis of neuromelanin. Mech Ageing Dev 63:157–163

Knoll J, Yen TT, Miklya I (1994) Sexually low performing male rats dies earlier than their high performing peers and (−)deprenyl treatment eliminates this difference. Life Sci 54:1047–1057

Knoll J, Miklya I, Knoll B, Markó R, Kelemen K (1996a) (−)Deprenyl and (−)1-phenyl-2-propylaminopentane, [(−)PPAP], act primarily as potent stimulants of action potential-transmitter release coupling in the catecholaminergic neurons. Life Sci 58: 817–827

Knoll J, Knoll B, Miklya I (1996b) High performing rats are more sensitive toward catecholaminergic activity enhancer (CAE) compounds than their low performing peers. Life Sci 58:945–952

Knoll J, Miklya I, Knoll B, Markó R, Rácz D (1996c) Phenylethylamine and tyramine are mixed-acting sympathomimetic amines in the brain. Life Sci 58:2101–2114

Knoll J, Yoneda F, Knoll B, Ohde H, Miklya I (1999) (−)1-(Benzofuran-2-yl)-2-propylaminopentane, [(−)BPAP], a selective enhancer of the impulse propagation mediated release of catecholamines and serotonin in the brain. Br J Pharm 128:1723–1732

Knoll J, Miklya I, Knoll B, Dalló J (2000) Sexual hormones terminate in the rat the significantly enhanced catecholaminergic/serotoninergic tone in the brain characteristic to the post-weaning period. Life Sci 67:765–773

Knoll J, Miklya I, Knoll B (2002a) Stimulation of the catecholaminergic and serotoninergic neurons in the rat brain by R-(−)-1-(benzofuran-2-yl)-2-propylaminopentane, (−)-BPAP. Life Sci 71:2137–2144

Knoll J, Miklya I, Knoll B, Yasusa T, Shimazu S, Yoneda F (2002b) 1-(Benzofuran-2-yl)-2-(3,3,3-trifluoropropyl)aminopentane HCl, 3-F-BPAP, antagonizes the enhancer effect of (−)-BPAP in the shuttle box and leaves the effect of (−)-deprenyl unchanged. Life Sci 71:1975–1984

Koffka K (1935) Principles of gestalt psychology. Harcourt, Brace, New York

Köhler W (1947) Gestalt psychology. Liveright, New York

Kuhn R (1957) Über die Behandlung depressiver Zustände mit einem Imonodibenzilderivat. Schweitz Med Wschr 36:1135–1139

Kuhn W, Muller T (1996) The clinical potential of deprenyl in neurological and psychiatric disorders. J Neural Transm Suppl 48:85–93

Larsen JP, Boas J, Erdal JE (1999) Does selegiline modify the progression of early Parkinson's disease? Results from a five-year study. The Norwegian-Danish Study Group. Eur J Neurol 6:539–547

Lees AJ (1991) Selegiline hydrochloride and cognition. Acta Neurol Scand Suppl 136:91–94

Lees AJ (1995) Comparison of therapeutic effects and mortality data of levodopa and levodopa combined with selegiline in patients with early, mild Parkinson's disease. Br Med J 311:1602–1607

Libet B (1973) Electrical stimulation of cortex in human subjects, and conscious memory aspects. In: Iggo A (ed) Handbook of sensory physiology, vol 2. Springer-Verlag, Berlin, Heidelberg, New York, pp 743–790

Levene M, Roberts P (eds) (1999) Massacre in history. Berghahn Books, New York

Lockhart BP, Lestage PJ (2003) Cognition enhancing or neuroprotective compounds for the treatment of cognitive disorders: why? when? which? Exp Gerontol 38:119–128

Lorente de No R (1935) Electrical excitability of motoneurones. J Cell Comp Physiol 7:46–71

Mann JJ, Gershon S (1980) A selective monoamine oxidase-B inhibitor in endogenous depression. Life Sci 26:877–882

Mantle TJ, Garrett NJ, Tipton KF (1976) The development of monoamine oxidase in rat liver and brain. FEBS Lett 64:227–230

Martin C (1977) Sexual activity in the aging male. In: Money J, Musaph H (eds) Handbook of sexology. Elsevier, Amsterdam, pp 813–824

Martini E, Pataky I, Szilágyi K, Venter V (1987) Brief information on an early phase-II study with (−)deprenyl in demented patients. Pharmacopsychiatry 20:256–257

Maruo J, Yoshida A, Shimohira I, Matsuno K, Mita S, Ueda H (2000) Binding of [^{35}S] GTPγS stimulated by (+)pentazocine, sigma receptor agonist, is abundant in the guinea pig spleen. Life Sci 67:599–603

McGeer EG, McGeer PL, Wada JK (1971) Distribution of tyrosine hydroxylase in human and animal brain. J Neurochem 18:1647–1658

McGrath PJ, Stewart JW, Harrison W, Wagner S, Nunes EN, Quitkin FM (1989) A placebo-controlled trial of L-deprenyl in atypical depression. Pschopharmacol Bull 25:63–67

Mendlewicz J, Youdim MB (1983) L-Deprenil, a selective monoamine oxidase type B inhibitor, in the treatment of depression: a double blind evaluation. Br J Psychiatry 142:508–511

Miklya I, Knoll J (2003) Analysis of the effect of (−)-BPAP, a selective enhancer of the impulse propagation mediated release of catecholamines and serotonin in the brain. Life Sci 72:2915–2921

Miklya I, Knoll B, Knoll J (2003a) A pharmacological analysis elucidating why, in contrast to (−)-deprenyl (selegiline) α-tocopherol was ineffective in the DATATOP study. Life Sci 72:2641–2648

Miklya I, Knoll B, Knoll J (2003b) An HPLC tracing of the enhancer regulation in selected discrete brain areas of food deprived rats. Life Sci 72:2923–2930

Milgram MW, Racine RJ, Nellis P, Mendoca A, Ivy GO (1990) Maintenance on L-(−)deprenyl prolongs life in aged male rats. Life Sci 47:415–420

Mink JW, Thach WT (1993) Basal ganglia intrinsic circuits and their role in behavior. Curr Opinion in Neurobiol 3:950–957

Miyoshi K (2001) Parkinson's disease. Nippon Rinsho 59:1570–1573

Monteverde A, Gnemmi P, Rossi F, Monteverde A, Finali GC (1990) Selegiline in the treatment of mild to moderate Alzheimer-type dementia. Clin Ther 12:315–322

Morishima-Kawashima M, Iharra Y (2002) Alzheimer's disease: β-amyloid protein and tau. J Neurosci Res 70:392–401

Moruzzi G, Magoun HW (1949) Brain stem reticular formation and activation of the EEG. Electroenceph Clin Neurophysiol 1:455–473

Moss FA (1924) Study of animal drives. J Exp Psychol 7:165–185

Myttyla VV, Sotaniemi KA, Vourinen JA, Heinonen EH (1992) Selegiline as initial treatment in de novo parkinsonian patiens. Neurology 42:339–343

Nakajima T, Kakimoto Y, Sano I (1964) Formation of β-phenylethylamine in mammalian tissue and its effect on motor activity. J Pharm 143:319–325

Neve RL, Robakis NK (1998) Alzheimer's disease: a re-examination of the amyloid hypothesis. Trends in Neurosci 21:15–19

Nguyen TV, Juorio AV (1989) Binding sites for brain trace amines. Cell Mol Neurobiol 9:297–311

Nguyen TV, Paterson IA, Juorio AV, Greenshow AJ, Boulton AA (1989) Tryptamine receptors: neurochemical and electrophysiological evidence for postsynaptic and functional binding sites. Brain Res 476:85–93

Nies A, Robinson DS, Davis JM, Ravaris CL (1973) Changes in monoamine oxidase with aging, In: Eisdorfer C, Fann WE (eds) Psychopharmacology of aging (Advances in behavioral biology). Plenum Press, New York, pp 41–54

Nussbaum RL, Ellis CE (2003) Alzheimer's disease and Parkinson's disease. N Engl J Med 348:1356–1364

Ohta K, Ohta M, Mizuta I, Fujinami A, Shimazu S, Sato N, Yoneda F, Hayashi K, Kuno S (2002) The novel catecholaminergic and serotonergic activity enhancer R-(−)-1-(benzofuran-2-yl)-2-propylaminopentane up-regulates neurotrophic factor synthesis in mouse astrocytes. Neurosci Lett 328: 205–208

Oka T, Yasusa T, Ando T, Watanabe M, Yoneda F, Ishida T, Knoll J (2001) Enantioselective synthesis and absolute configuration of (−)1-(benzofuran-2-yl)-2-propylaminopentane, (−)-BPAP, a highly potent and selective catecholaminergic activity enhancer. Bioorg Med Chem 9:1213–1219

Olanow CW, Godbold JH, Koller W (1996) Effect of adding selegiline to levodopa in early, mild Parkinson's disease. Patients taking selegiline may have received more levodopa than necessary. Br Med J 312:702–703

Parkinson Study Group (1989) Effect of (−)deprenyl on the progression disability in early Parkinson's disease. New Engl J Med 321:1364–1371

Parkinson Study Group (1993) Effect to tocopherol and (−)deprenyl on the progression of disability in early Parkinson's disease. New Engl J Med 328: 176–183

Parkinson Study Group (1996) Impact of deprenyl and tocopherol treatment of Parkinson's disease in DATATOP patients requiring levodopa. Ann Neurol 39:37–45

Pavlov IP (1955) Pawlowsche Mittwochkolloquien. Akademie Verlag, Berlin, p 492

Penfield W (1955) The permanent record of the stream of consciousness. Acta Psychologica 11:47–69

Phillips PEM, Stuber GD, Heien MLAV, Wightman RM, Carelli RM (2003) Subsecond dopamine release promotes cocaine seeking. Nature 422:614–617

Ponto LL, Schultz SK (2003) Ginkgo biloba extract: review of CNS effects. Ann Clin Psychiatry 15:109–119

Popper RK, Eccles JC (1977) The self and its brain. An argument for interactionism. Springer International, Berlin

Premont RT, Gainetdinov RR, Caron MG (2001) Following the trace of elusive amines. Proc Natl Acad Sci USA 98:9474–9475

Quitkin RT, Liebowitz MR, Stewart JW, McGrath PJ, Harrison W, Rabkin JG, Markowitz J, Davis SO (1984) L-Deprenyl in atypical depression. Arch Gen Psychiatry 41:777–781

Rashid MH, Matsumoto T, Mizuno K, Watanabe M, Sato N, Yoneda F, Ueda H (2001) Nociceptive responses by deprenyl derivative, (−)BPAP through metabotropic sigma receptor. Pharmacol Rev Com 11:335–342

Richards JG, Lorez HP, Tranzer JP (1973) Indolealklylamine nerve terminals in cerebral ventricles, identification by electron microscopy and fluorescence histochemistry. Brain Res 57:277–288

Richter CP (1927) Animal behavior and internal drives. Quart Rev Biol 2: 307–343

Riederer P, Wuketich S (1976) Time course of nigrostriatal degeneration in Parkinson's disease. J Neural Transm 38:277–301

Rinne JO, Röyttä M, Paljärvi L, Rummukainen J, Rinne UK (1991) Selegiline (deprenyl) treatment and death of nigral neurons in Parkinson's disease. Neurology 41:859–861

Ritter JL, Alexander B (1997) Retrospective study of selegiline-antidepressant drug interactions and a review of the literature. Ann Clin Psychiarty 9:7–13

Robinson DS, Davis JM, Nies A, Ravaris CL, Sylwester D (1971) Relation of sex and aging to monoamine oxidase activity of human brain, plasma and platelets. Arch Gen Psychiatry 24:536–539

Robinson DS, Davis JM, Nies A, Colburne JR, Runney WE, Shaw DM (1972) Aging, monoamines, and monoamine oxidase levels. The Lancet i:290–291

Saavedra JM (1974) Enzymatic isotopic assay for and presence of beta-phyenylethylamine in brain. J Neurochem 22:211–216

Saavedra JM (1989) Catecholamines II. In: Trendelenburg U, Weiner N (eds) Handbook of experimental pharmacology. Springer-Verlag, Berlin Heidelberg New York, pp 181–201

Sabelli HC, Mosnaim AD (1974) Phenylethylamine hypothesis of affective behavior. Am J Psychiat 131:695–699

Sabelli HC, Javaid JI (1995) Phenylethylamine modulation of affect: therapeutic and diagnostic implication. J Neurophsychiatry Clin Neurosci 7:6–14

Sabelli HC, Fawcett J, Gusovsky F, Javaid JI, Wynn P, Edwards H, Jeffriess H, Kravitz HJ (1986) Clinical studies on the phenylethylamine hypothesis of affective disorder: urine and blood phenylacetic acid and phenylalanine dietary supplements. Clin Psychiatry 47:66–70

Sandler M, Glover V, Ashford A, Stern GM (1978) Absence of "cheese effect" during deprenyl therapy: some recent studies. J Neural Transm 43:209–215

Sano M, Ernesto C, Klauber MR (1996) Rationale and design of a multicenter study of selegiline and α-tocopherol in the treatment of Alzheimer disease using novel clinical outcomes. Alzheimer Dis Assoc Disord 10:132–140

Sano M, Ernesto C, Thomas RG, Klauber MR, Schafer K, Grundman M, Woodbury P, Growdon J, Cotman CW, Pfeiffer E, Schneider LS, Thal LJ (1997)

A controlled trial of selegiline, alpha-tocopherol, or both as treatment for Alzheimer's disease. N Engl J Med 336:1216–1222

Satoi M, Matsuishi T, Yamada S, Yamashita Y, Ohtaki E, Mori K, Kiikonen R, Kato H, Percy AK (2000) Decreased cerebrosbinal fluid levels of beta-phenylethylamine in patients with Rett snydrome. Ann Neurol 47:801–803

Schumacher M, Weil-Engerer S, Liere P, Robert F, Franklin RJ, Garcia-Segura LM, Lambert JJ, Mayo W, Melcangi RC, Parducz A, Suter U, Carelli C, Baulieu EE, Akwa Y (2003) Steroid hormones and neurosteroids in normal and pathological aging of the nervous system. Prog Neurobiol 71:3–29

Selkoe DJ (2001) Alzheimer's disease: genes, proteins, and therapy. Physiol Rev 81:741–766

Shih JC (1979) Monoamine oxidase in aging human brain. In: Singer TP, Korff RW, Murphy DL (eds) Monoamine oxidase: structure, function and altered functions. Academic Press, New York, pp 413–421

Shimazu S, Miklya I (2004) Pharmacological studies with endogenous enhancer substances: β-phenylethyamine, tryptamine, and their synthetic derivatives. Progr Neuro-Psychopharmacol Biol Psychiatry 28:421–427

Shimazu S, Takahata K, Katsuki H, Tsunekawa H, Tanigawa A, Yoneda F, Knoll J, Akaike A (2001) (−)-1-(Benzofuran-2-yl)-2-proplyaminopentane enhances locomotor activity in rats due to its ability to induce dopamine release. Eur J Pharm 421:181–189

Shimazu S, Tanigawa A, Sato N, Yoneda F, Hayashi K, Knoll J (2003) Enhancer substances: Selegiline and R-(−)-1-(benzofuran-2-yl)-2-propylaminopentane, [(−)-BPAP] enhance the neurotrophic factor synthesis on cultured mouse astrocytes. Life Sci 72:2785–2792

Skinner BF (1938) The behavior of organisms. Appleton Century Crofts, New York

Stamford JA, Justice JB Jr (1996) Probing brain chemistry. Anal Chem 68:359A–363A

Standaert DG, Young AB (1996) Treatment of central nervous system degenerative disorders. In: Hardman JG, Limbird LE (eds) Goodman & Gilman's. The pharmacological basis of therapeutics, 9th edn. McGraw-Hill, New York, pp 503–519

Strolin Benedetti M, Keane PE (1980) Differential changes in monoamine oxidase A and B activity in the aging brain. J Neurochem 35:1026–1032

Student AK, Edwards DJ (1977) Subcellular localization of types A and B monoamine oxidase in rat brain. Biochem Pharm 26:2337–2342

Tanner CM, Goldmann SM (1996) Epidemiology of Parkinson's disease. Neurol Clin 14:317–335

Tariot PN, Cohen RM, Sunderland T, Newhouse PA, Yount D, Mellow AM (1987) L-(−)Deprenyl in Alzheimer's disease. Arch Gen Psychiatry 44:427–433

Tennyson V, Heikkila R, Mytilineou C, Coté L, Cohen G (1974) 5-Hydroxydopamine 'tagged' neuronal boutons in rabbit neostriatum: interrelationship between vesicles and axonal membrane. Brain Res 82:341–348

Tetrud JW, Langston JW (1989) The effect of (−)deprenyl (selegiline) on the natural history of Parkinson's disease. Science 245:519–522

Thomas T (2000) Monoamine oxidase-B inhibitors in the treatment of Alzheimer's disease. Neurobiol Aging 21:343–348

Thorndike EL (1898) Animal intelligence: An experimental study on the associative processes in animals. Psychol Rev 2(8) (monograph suppl)

Thorndike EL (1911) Animal intelligence. Macmillan, New York

Thorndike EL (1940) Psychology of wants, interests and attitudes. Macmillan, New York

Tokuyama S, Hirata K, Yoshida A, Maruo J, Matsuno K, Mita S, Ueda H (1999) Selective coupling of mouse brain metabotropic sigma receptor with recombinant G_{i1}. Neurosci Lett 268:85–88

Tolman EC (1932) Purposive behaviour in animals and in man. Appleton Century Crofts, New York

Tom T, Cummings JL (1998) Depression in Parkinson's disease. Pharmacological characteristics and treatment. Drugs Aging 12:55–74

Tóth V, Kummert M, Sugár J, Knoll J (1992) A procedure for measuring neuromelanin in neurocytes by a TV-image analyser. Mech Ageing Dev 63: 215–221

Tringer L, Haits G, Varga E (1971) The effect of (−)E-250, (−)L-phenyl-isopropylmethyl- propinyl-amine HCl, in depression. In: Leszkovszky G (ed) V. Conferentia Hungarica pro Therapia et Investigatione in Pharmacologia. Akadémiai Kiadó, Budapest, pp 111–114

Uchtomsky AA (1945) Cited by Knoll J (1969) The theory of active reflexes. Hungarian Academy of Sciences, Budapest, Hafner Publishing Company, New York, p 75

Ueda H, Inoue M (2000) In vivo signal transduction of nociceptive responses by kyotorphin (tyrosis-arginin) through $G\alpha_i$- and inositol triphosphate-mediated Ca^{2+} influx. Molec Pharm 57:108–115

Usdin E, Sandler M (eds) (1976) Trace amines and the brain. Marcel Dekker, New York

Varga E (1965) Vorläufiger Bericht über die Wirkung des Präparats E-250 (Phenyl-Isopropyl-Methyl-Propinylamine-Chlorhydrat) In: Dumbovich B (ed) III. Conferentia Hungarica pro Therapia et Investigatione in Pharmacologia. Akadémiai Kiadó (Hungarian Academy of Sciences), Budapest, pp 197–201

Varga E, Tringer L (1967) Clinical trial of a new type of promptly acting psychoenergetic agent (phenyl-isopropylmethyl-propinylamine HCl, E-250). Acta Med Acad Sci Hung 23:289–295

Vizuete ML, Steffen V, Ayala A, Cano J, Machado A (1993) Protective effect of deprenyl against 1-methyl-4-phenylpiridinium neurotoxicity in rat striatum. Neurosci Lett 152:113–116

Walker SE, Shulman KI, Tailor SA, Gardner D (1996) Tyramine content of previously restricted foods in monoamine oxidase inhibitor diets. J Clin Psychopharmacol 16:383–388

Wilner J, LeFevre HF, Costa E (1974) Assay by multiple ion detection of phenylethylamine and phenylethanolamine in rat brain. J Neurochem 23: 857–859

Wu RM, Chiuech CC, Pert A, Murphy DL (1993) Apparent antioxidant effect of l-deprenyl on hydroxyl radical formation and nigral injury elicited by MPP$^+$ in vivo. Eur J Pharmacol 243:241–247

Yoneda F, Moto T, Sakae M, Ohde H, Knoll B, Miklya I, Knoll J (2001) Structure-activity studies leading to (−)1-(benzofuran-2-yl)-2-propylaminopentane, (−)BPAP, a highly potent, selective enhancer of the impulse propagation mediated release of catecholamines and serotonin in the brain. Bioorg Med Chem 9:1197–1212

Youdim MB (1980) Monoamine oxidase inhibitors as anti-depressant drugs and as adjunct to L-dopa therapy of Parkinson's disease. J Neural Transm Suppl 16:157–161

Zeller EA, Barsky J (1952) In vivo inhibition of liver and brain monoamine oxidase by 1-isonicotinyl-2-isopropylhydrazine. Proc Soc Exp Biol Med 81:459–468

Zeller EA, Barsky J, Fouts JE, Kirchheimer WF, Van Orden LS (1952) Influence of isonicotinic acid hydrazide (INH) and 1-isonicotinic 2-isopropylhydrazide (IIH) on bacterial and mammalian enzymes. Experientia 8:349–350

Zesiewicz TA, Gold M, Chari G, Hauser RA (1999) Current issues in depression in Parkinson's disease. Am J Geriat Psychiat 7:110–118

Index of Names

Page numbers in *italics* refer to the References.

A

Adham N, *see* Borowsky B
Agosti C, *see* Borroni B
Ajika K, 14, *147*
Akkawi N, *see* Borroni B
Akwa Y, *see* Schumacher M
Albin RL, 89, *147*
Alexander B, 88, *see* Ritter JL
Allain H, 29, 91, *147*
Alonso G, *see* Calas A
Alzheimer A, 92
Amara SG, *see* Bunzow JR
Ambrózi L, *see* Birkmayer W
Amiel HF, 122
Amsterdam JD, 88, *147, see* Bodkin JA
Ando T, *see* Oka T
Archetti S, *see* Borroni B
Arcimboldo G, 131
Arnauld E, *see* Calas A
Artimishyn R, *see* Borowsky B
Arttamangkul S, *see* Bunzow JR
Ashford A, *see* Sandler M
Auden WH, 99
Awouters HF, *see* Janssen PA
Ayala A, *see* Vizuete ML

B

Bach JS, 52
Ban TA, 2, *147*
Barsky J, 28, *see* Zeller EA
Bartov O, 134, *147*
Baudelaire C, 102
Baulieu EE, *see* Schumacher M
Beethoven L van, 114–115
Belliard S, *see* Campion D
Beritov IC, 5, *147*
Bertler A, 75, *147*
Besson NJ, *see* Calas A
Birkmayer W, 29, 88, *147*
Birks J, 92, *148*
Blackwell B, 28, *148*
Boas J, *see* Larsen JP
Bodkin JA, 88, *148*
Bonini JA, *see* Borowsky B
Bookheimer SY, 94, *148*
Borowsky B, 48, 49, 74, *148*
Borroni B, 94, *148*
Bosch H, 130
Boulton AA, 73–74, *148, see* Davis BA, *see* Nguyen TV
Bovet D, 51, *see* Kelemen K
Boyle N, *see* Borowsky B
Brambilla C, *see* Borroni B
Branchek TA, *see* Borowsky B
Braque G, 126
Bunzow JR, 48, 74, *148*

C

Caimi L, *see* Borroni B

Calas A, 14, *148*
Caltagirone C, *see* Borroni B
Campi N, 92, *148*
Campion D, 93, *148*
Cano J, *see* Vizuete ML
Carelli C, *see* Schumacher M
Carelli RM, *see* Phillips PEM
Carlsson A, 74, *148*
Caron MG, *see* Premont RT
Carrillo MC, 86, *149*, *see* Kitani K
Cauchy C, *see* Calas A
Cézanne P, 125–126
Chan-Palay V, 14, *149*
Chari G, *see* Zesiewicz TA
Charlett A, *see* Dobbs SM
Cheramy A, *see* Calas A
Chiuech CC, *see* Wu RM
Cohen A, *see* Cohen GP
Cohen G, *see* Tennyson V
Cohen GP, 86, *149*
Cohen MS, *see* Bookheimer SY
Cohen RM, *see* Tariot PN
Colburne JR, *see* Robinson DS
Commenges D, *see* Helmer C
Connick JH, 49, *149*
Costa E, *see* Wilner J
Coté L, *see* Tennyson V
Cotman CW, *see* Sano M
Crane GE, 28, *149*
Cummings JL, 88, *see* Tom T

D

Dabrowsky M, 55, *149*
Dalló J, 84, *149*, *see* Knoll J
Darland T, *see* Bunzow JR
Dartigues JF, *see* Helmer C
Davis BA, 73, *149*
Davis JM, *see* Nies A, *see* Robinson DS
Davis SO, *see* Quitkin RT
DeLanney LE, *see* Finnegan KT
Descarries L, 14, *149*

Di Luca M, *see* Borroni B
Diderot D, 124
Dobbs RJ, *see* Dobbs SM
Dobbs SM, 91, *149*
Downer RGH, *see* Boulton AA
Dubois B, *see* Campion D
Dumanchin C, *see* Campion D
Durkin MM, *see* Borowsky B

E

Eccles JC, 6, 139, *see* Popper RK
Eckert B, 74, *149*
Ecseri Z, *see* Knoll J
Edwards DJ, 74, *see* Student AK
Edwards H, *see* Sabelli HC
Einstein A, 1
Ellis CE, 93, *see* Nussbaum RL
Elsworth JD, 29, *150*
Ensor J, 122
Erdal JE, *see* Larsen JP
Ernesto C, *see* Sano M
Etminan M, 92, *150*
Eyck van Brothers, 23, 130

F

Falsaperla A, 93, *150*
Fawcett J, *see* Sabelli HC
Finali GC, *see* Monteverde A
Finnegan KT, 86, *150*
Fischer E, 73, *150*
Fischer E, 23
Flicker L, 92, *see* Birks J
Fouts JE, *see* Zeller EA
Fowler CJ, 74, *150*
Fox PK, *see* Connick JH
France L, *see* Connick JH
Franklin RJ, *see* Schumacher M
Freud S, 1, 116–117, 143
Fujinami A, *see* Ohta K
Fujita R, *see* Hamabe W

G

Gainetdinov RR, see Premont RT
Garcia-Segura LM, see Schumacher M
Gardner D, see Walker SE
Garrett NJ, see Mantle TJ
Gauguin P, 140
Gerald C, see Borowsky B
Gershon S, 88, see Mann JJ
Gill S, see Etminan M
Glover V, see Elsworth JD, see Sandler M
Glowinsky J, see Calas A
Gnemmi P, see Monteverde A
Godbold JH, see Olanow CW
Goethe JW, 122, 138
Gold M, see Zesiewicz TA
Goldmann SM, see Tanner CM
Gottfries CG, see Eckert B
Gougnard J, see Allain H
Grandy DK, see Bunzow JR
Greco El, 125
Greenshow AJ, 73, *150*, see Nguyen TV
Growdon J, see Sano M
Grundman M, 92, *150*, see Sano M
Gusovsky F, see Sabelli HC

H

Haits G, see Tringer L
Hamabe W, 41, 49, *150*
Händel GF, 97
Hanlon G, see Connick JH
Hannequin D, see Campion D
Harrison LM, see Bunzow JR
Harrison W, see McGrath PJ, see Quitkin RT
Hárs V, see Birkmayer W
Hársing RG, 86, *150*
Hauger RL, 74, *150*
Hauser RA, see Zesiewicz TA
Hayashi K, see Ohta K

Haydn FJ, 97
Healy D, 2, see Ban TA, *150, 151*
Heien MLAV, see Phillips PEM
Heikkila R, see Tennyson V
Heinonen EH, see Myttyla VV
Heller B, see Fischer E
Helmer C, 93, *151*
Hilgard ER, 1, *151*
Hirata K, see Tokuyama S
Hitler A, 135
Hökfelt T, 14, see Ajika K, *151*
Hornykiewicz O, 29, see Birkmayer W
Hy LX, 93, *151*

I

Iacopino V, 134, *151*
Iharra Y, 92, see Morishima-Kawashima M
Inoue M, see Ueda H
Irvin I, see Finnegan KT
Ishida T, see Oka T
Isobe K, see Kitani K
Ivy GO, see Carrillo MC

J

Janssen PA, 74, *151*
Javaid JI, 73, see Sabelli HC
Jawlensky A, 126–127
Jeffriess H, see Sabelli HC
Johnston JP, 29, *151*
Joly P, see Helmer C
Jones KA, see Borowsky B
Juhász Gy, 124
Juorio AV, 74, see Boulton AA, see Nguyen TV
Justice JBjr, 114, see Stamford JA

K

Kahlo F, 127
Kanai S, see Carrillo MC

Kandinsky W, 55, 126
Kato H, *see* Satoi M
Katsuki H, *see* Shimazu S
Kaufman S, 134, *151*
Keane PE, 74, *see* Strolin Benedetti M
Kelemen K, 51, *151*, *see* Knoll J
Keller DM, *see* Hy LX
Kennedy JL, *see* Bunzow JR
Kiikonen R, *see* Satoi M
King David, 136
Kirchheimer WF, *see* Zeller EA
Kitani K, 92, *see* Carrillo MC, *151*
Klauber MR, *see* Sano M
Klee P, 126
Kline NS, 28, *151*
Knoll B, 50, *151*, *see* Knoll J, *see* Miklya I, *see* Yoneda F
Knoll J, 2, 6, 8–9, 13–14, 17–20, 25–30, 32–41, 43–44, 47–49, 51, 53, 65–71, 74–75, 77–82, 84, 86–87, 89, 91–92, 94, 95, 131, *see* Birkmayer W, *see* Dalló J, *see* Hársing RG, *see* Kelemen K, *151-155*, *see* Miklya I, *see* Oka T, *see* Shimazu S, *see* Tóth V, *see* Yoneda F
Knorring L, *see* Eckert B
Koffka K, 1, 5, *155*
Köhler W, 1, *155*
Koller W, *see* Olanow CW
Kouranova E, *see* Borowsky B
Kravitz HJ, *see* Sabelli HC
Kubin A, 124
Kuhn R, 28, *155*
Kuhn W, 88, *155*
Kumar P, *see* Elsworth JD
Kummert M, *see* Knoll J, *see* Tóth V
Kuno S, *see* Ohta K

L

Lakhlani PP, *see* Borowsky B
Lambert JJ, *see* Schumacher M
Langston JW, 29–31, *see* Finnegan

KT, *see* Tetrud JW
Lapierre Y, *see* Descarries L
Larsen JP, 29, 91, *155*
Lees AJ, 88, 91, *see* Elsworth JD, *155*
LeFevre HF, *see* Wilner J
Léger L, *see* Descarries L
Leist C, *see* Cohen GP
Lekka N, *see* Dalló J
Leonardo da Vinci, 125, 127, 132
Lestage PJ, 94, *see* Lockhart BP
Letenneur L, *see* Helmer C
Levene M, 134, *155*
Leysen JE, *see* Janssen PA
Libet B, 58–59, 120, *155*
Lichtblau H, *see* Borowsky B
Liebowitz MR, *see* Quitkin RT
Liere P, *see* Schumacher M
Lockhart BP, 94, *155*
Lodovico il Moro (Sforza), 132
Longo VG, *see* Kelemen K
Lorente de No R, 5, *155*
Lorez HP, *see* Richards JG
Lully J-B, 97
Lysippos, 132

M

Machado A, *see* Vizuete ML
Maehara K, *see* Kitani K
Magenis RE, *see* Bunzow JR
Magoun HW, 1, *see* Moruzzi G
Magyar K, 29, *see* Hársing RG, *see* Knoll J
Mallory GL, 112
Mann JJ, 88, *155*
Mantle TJ, 74, *155*
Marcusson J, *see* Fowler CJ
Markó R, *see* Knoll J
Markowitz J, *see* Quitkin RT
Martin C, 76–77, *156*
Martini E, 92, *156*
Marton V, *see* Birkmayer W
Maruo J, 49, *156*, *see* Tokuyama S

Matisse H, 126
Matsuishi T, *see* Satoi M
Matsumoto T, *see* Rashid MH
Matsuno K, *see* Maruo J, *see* Tokuyama S
Mayo W, *see* Schumacher M
Mazziotta JC, *see* Bookheimer SY
McGeer EG, 75, *156*
McGeer PL, *see* McGeer EG
McGrath PJ, 88, *156*, *see* Quitkin RT
Megens AA, *see* Janssen PA
Melcangi RC, *see* Schumacher M
Mellow AM, *see* Tariot PN
Mendlewicz J, 88, *156*
Miklya I, 15, 18–20, 25, 32–38, 40, 65–69, 74, 89, 131, *see* Knoll J, *156*, *see* Shimazu S, *see* Yoneda F
Minami C, *see* Kitani K
Mink JW, 89, *156*
Miró J, 126
Miró AH, *see* Fischer E
Mita S, *see* Maruo J, *see* Tokuyama S
Mitileneou MD, *see* Cohen GP
Miyoshi K, 88, *156*
Mizuno K, *see* Rashid MH
Mizuta I, *see* Ohta K
Mohamed, 136
Monici Preti PA, *see* Falsaperla A
Monteverde A, 93, *156*
Monteverde A, *see* Monteverde A
Mori K, *see* Satoi M
Morishima-Kawashima M, 92, *156*
Moruzzi G, 1, *156*
Mosnaim AD, 73, *see* Sabelli HC
Moss FA, 1, *156*
Mozart L, 97
Mozart WA, 97–98, 111, 121
Muller T, 88, *see* Kuhn W
Münter G, 55
Murphy DL, *see* Wu RM
Mytilineou C, *see* Tennyson V
Myttyla VV, 29, 91, *156*

N

Naukirek HC, *see* Allain H
Neve RL, 92, *156*
Newhouse PA, *see* Tariot PN
Nguyen TV, 74, *157*
Nicholson CD, *see* Connick JH
Nies A, 74, *157*, *see* Robinson DS
Nietzsche F, 127
Nievel J, *see* Knoll J
Nokubo M, *see* Carrillo MC
Nunes EN, *see* McGrath PJ
Nussbaum RL, 93, *157*

O

Ochoa FY, *see* Borowsky B
Ogozalek KL, *see* Borowsky B
Ohde H, *see* Knoll J, *see* Yoneda F
Ohta K, 41–42, *157*
Ohta M, *see* Ohta K
Ohtaki E, *see* Satoi M
Oka T, 38, *157*
Olanow CW, 91, *157*
Oliani C, *see* Falsaperla A
Olson SB, *see* Bunzow JR
Oreland L, *see* Eckert B, *see* Fowler CJ

P

Padovani A, *see* Borroni B
Paljärvi L, *see* Rinne JO
Párducz A, *see* Schumacher M
Parkinson Study Group, 29–31, 85, 91, *157*
Pasik B, *see* Cohen GP
Pasumamula S, *see* Bunzow JR
Pataky I, *see* Martini E
Paterson IA, *see* Nguyen TV
Pathirana S, *see* Borowsky B
Paul SM, *see* Hauger RL
Pavlov IP, 1, 6, 118, *157*
Peel M, 134, *see* Iacopino V
Penfield W, 117, *157*

Penney JB, see Albin RL
Percy AK, see Satoi M
Pericak-Vance MA, see Bookheimer SY
Pert A, see Wu RM
Pfeiffer E, see Sano M
Phillips PEM, 114, *157*
Phuapradit P, see Elsworth JD
Picasso P, 126
Ponto LL, 92, *157*
Popper RK, 6, 139, *157*
Premont RT, 73–74, *157*
Pu X, see Borowsky B
Puel M, see Campion D

Q

Quigley DI, see Bunzow JR
Quitkin RT, 88, see McGrath PJ, *158*

R

Rabkin JG, see Quitkin RT
Rácz D, see Knoll J
Raddatz R, see Borowsky B
Rashid MH, 49, *158*
Ravaris CL, see Nies A, see Robinson DS
Ravel M, 21
Reggiani H, see Fischer E
Reynolds GP, see Elsworth JD
Richards JG, 14, *158*
Richter CP, 1, *158*
Riederer P, 75, see Birkmayer W, *158*
Rinne JO, 87, *158*
Rinne UK, see Rinne JO
Ritter JL, 88, *158*
Robakis NK, 92, see Neve RL
Robert F, see Schumacher M
Roberts J, see Connick JH
Roberts P, 134, see Levene M
Robinson DS, 74, see Nies A, *158*
Rossi F, see Monteverde A
Röyttä M, see Rinne JO
Rummukainen J, see Rinne JO
Runney WE, see Robinson DS

S

Saavedra JM, 74, *158*
Sabelli HC, 73, *158*
Saint Phalle de Niki, 124
Samii A, see Etminan M
Sandler M, 29, 74, see Elsworth JD, *158*, see Usdin E
Sano M, 92–93, *158*
Sato N, see Ohta K, see Rashid MH
Sato Y, see Carrillo MC
Satoi M, 74, *159*
Saunders AM, see Bookheimer SY
Scarzella L, see Campi N
Schafer K, see Sano M
Schneider LS, see Sano M
Schultz SK, 92, see Ponto LL
Schumacher M, 92, *159*
Selkoe DJ, 92, *159*
Shakespeare W, 124
Shaw DM, see Robinson DS
Shaw KM, see Elsworth JD
Shih JC, 74, *159*
Shimazu S, 41–42, 74, see Ohta K, *159*
Shimohira I, see Maruo J
Shorter E, see Ban TA
Shulman KI, see Walker SE
Skinner BF, 1, *159*
Skolnick P, see Hauger RL
Skratt JJ, see Finnegan KT
Small GW, see Bookheimer SY
Somogyi G, see Knoll J
Sonders MS, see Bunzow JR
Sotaniemi KA, see Myttyla VV
Spatz H, see Fischer E
Stamford JA, 114, *159*
Standaert DG, *159*
Steffen V, see Vizuete ML
Stern GM, see Elsworth JD, see Sandler M

Stewart JW, *see* McGrath PJ, *see* Quitkin RT
Strojwas MH, *see* Bookheimer SY
Strolin Benedetti M, 74, *159*
Stuber GD, *see* Phillips PEM
Student AK, 74, *159*
Suchland KL, *see* Bunzow JR
Sugár J, *see* Knoll J, *see* Tóth V
Sunderland T, *see* Tariot PN
Suter U, *see* Schumacher M
Sylwester D, *see* Robinson DS
Szilágyi K, *see* Martini E

T

Tailor SA, *see* Walker SE
Takahata K, *see* Shimazu S
Tanigawa A, *see* Shimazu S
Tanner CM, 93, *159*
Tariot PN, 92, *159*
Tekes K, *see* Hársing RG
Tennyson V, 14, *160*
Tetrud JW, 29–31, *160*
Thach WT, 89, *see* Mink JW
Thal LJ, *see* Sano M
Thomas RG, *see* Sano M
Thomas T, 92, *160*
Thorndike EL, 1, *160*
Tímár J, *see* Knoll J
Tipton KF, *see* Mantle TJ
Todeschini GP, *see* Campi N
Tokuyama S, 49, *160*
Tolman EC, 1, *160*
Tom T, 88, *160*
Török Z, *see* Knoll J
Tóth V, 86, *see* Knoll J, *160*
Tranzer JP, *see* Richards JG
Tringer L, 88, *160*
Tsunekawa H, *see* Shimazu S

U

Uchtomsky AA, 5, *160*

Ueda H, 49, *see* Hamabe W, *see* Maruo J, *see* Rashid MH, *see* Tokuyama S, *160*
Usdin E, 74, *160*

V

Varga E, 88, *see* Tringer L, *160*
Venter V, *see* Martini E
Verrocchio, Andrea del (Andrea di Cione), 132
Vincent JD, *see* Calas A
Vizi ES, *see* Hársing RG, *see* Knoll J
Vizuete ML, 86, *161*

W

Wada JK, *see* McGeer EG
Wagner S, *see* McGrath PJ
Walker SE, 74, *161*
Watanabe M, *see* Oka T, *see* Rashid MH
Watkins KC, *see* Descarries L
Weil-Engerer S, *see* Schumacher M
Wightman RM, *see* Phillips PEM
Wilbert A, *see* Eckert B
Wilner J, 73, *161*
Winblad B, *see* Eckert B, *see* Fowler CJ
Woodbury P, *see* Sano M
Wright FL, 127
Wu RM, 86, *161*
Wuketich S, 75, *see* Riederer P
Wynn P, *see* Sabelli HC

Y

Yahr M, *see* Cohen GP
Yamada S, *see* Satoi M
Yamashita Y, *see* Satoi M
Yasar S, *see* Knoll J
Yasusa T, *see* Hamabe W, *see* Oka T

Yen TT, *see* Knoll J

Yoneda F, 38, 48, *see* Hamabe W, *see* Knoll J, *see* Ohta K, *see* Oka T, *see* Rashid MH, *see* Shimazu S, *161*

Yoshida A, *see* Hamabe W, *see* Maruo J, *see* Tokuyama S

Youdim MBH, 88, *see* Birkmayer W, *see* Mendlewicz J, *161*

Young AB, *see* Albin RL, *see* Standaert DG

Yount D, *see* Tariot PN

Z

Zeller EA, 28, *161*

Zesiewicz TA, 88, *161*

Zhang G, *see* Bunzow JR

Index of Subjects

A

acetyl-L-carnitine, 93
acquired drives, 2, 4–7, 11, 12, 18, 21, 25, 120, 129, 139, 141, 144
actinomycin D, 43
active focus (specific activation), 2, 6, 14, 52, 54, 111, 117, 127
affective state of consciousness, 5, 115, 119, 129, 142, 144
aging, 72–78
– chronological age, 72
– decrement of functions, 72
– downhill period of life, 63–66, 73–81
– external appearance, 72
– physiological age, 72
– uphill period of life, 63–66
Alzheimer's disease, 9, 71, 92–94, 145
– genetic risk factors, 94
– geographical differences in incidence, 94
– prevalence, 93
– sex differences in incidence, 94
amalgamation of order and chaos in the brain, 116, 118, 120
amphetamine, 18, 19, 28, 32, 36
β-amyloid cascade theory, 92
β-amyloid$_{1-40}$, 92
β-amyloid$_{1-42}$, 92
β-amyloid$_{25-35}$, 43, 94
anti-inflammatory drugs, 92
anti-semitism, 135, 136
antioxidant, 31
apperception of the self, 113
Arab–Israeli conflict, 135
ascending reticular activity system, 1
association, 1
astroglia, 42
attention deficit/hyperactive disorder, 74
Auschwitz, 2
Austria, 121

B

basal ganglia, 89, 90
BD 1063, sigma receptor antagonist, 49
boredom, 102–112
brain
– aging, 62–66, 73–78
– brain/self relation, 5
– irrational activity/chaos, 116, 117, 124, 128, 129, 134
– manipulability, 2, 3, 22, 132, 134, 137–140, 144
– rational activity/order, 116, 117, 124, 128, 129, 134
brain cells in culture, 40–47
– cortical neurons, 44
– glial cells, 42, 43
– hippocampal neurons, 43, 44
brain stem, 26, 27, 59, 68–72
bromocryptine, 38

C

Canary Islands, 135
catecholaminergic neurons, 8, 20, 66, 69
- dopaminergic neurons, 14, 68, 69
- noradrenergic neurons, 14, 68, 69
catecholamines, 14
- dopamine, 15, 20, 25–27, 32, 67–71, 114
- norepinephrine (noradrenaline), 15, 25–28, 32, 67–71
cheese effect, 28
cholinergic strial interneurons, 89
Christian world/Christianity, 135, 136
church, 124, 133, 138
clorgyline, 29, 38
cocaine, 114
coexistence of order and chaos in human brain, 113, 120
cognitive/volitional consciousness, 5, 114, 115, 129, 142, 144
concept, 6
conditioned avoidance response (CAR), 44, 46, 50, 56–60, 62
conscious perception, 5, 6, 58, 120, 129, 130
cortical neurons, belonging to
- Group 1, 5, 6, 51, 52, 113, 142
- Group 2, 5, 6, 51, 52, 113, 142
- Group 3, 5, 6, 51–53, 113, 115–117, 130, 142
- Group 4, 5, 6, 52–54, 113, 115–117, 120, 127, 130, 142
creative mind, 121, 127
cubist revolution, 126
cultural heritage, 132

D

DATATOP multicenter study of the Parkinson Study Group, 29–31, 85
depression, 28, 73, 88, 89
- atypical, 88, 89
- major, 88, 89
desmethylimipramine, 38
disestablishment of the church, 124, 138
Domestication/domesticated animals, 53
domestication/domesticated animals, 12, 17, 22
dominant focus, 5
downhill period of life, 8, 62, 145

E

E-250, *see* (−)-deprenyl
EEG, 8, 51, 64
emotion, 113, 114
emotional misbelieves, 134
engine of the brain, 64
England, 135
enhancer receptors, 47–49, 61
enhancer regulation, 2, 17, 20, 25, 27, 76
- age-related decline of regulation, 73–79, 81
- cortical enhancer regulation, 2, 5, 143
- dampened regulation after sexual maturity, 66–69
- enhanced regulation after weaning, 62–66
- mesencephalic enhancer regulation, 2, 8, 14, 19, 25, 39–42, 131, 145
  - nonspecific form, 39–42
  - specific form, 39–42
enhancer substances, 5, 14, 47–49, 119
- (−)-BPAP: R-(−)-1(benzofuran-2-yl)-2-propylaminopentane, 36–39, 48, 49, 61, 94, 95, 119, 145
  - enhancer effect, 37–39
  - peculiar dose/dependency of

Index of Subjects

enhancer effect, 39, 40
- cortical enhancer substances, 55–60, 62
- (–)-deprenyl (Selegiline), 27–36, 86–94, 145
  - absence of catacholamine-releasing property, 28, 29
  - antiaging effect, 62–66
  - antidepressant effect, 88, 89
  - E-250 – orginal of code name, 28
  - enhancer effect, 33–36
  - MAO-B inhibitory effect, 29–33
  - morphological evidence of antiaging effect, 86
  - potentiation of levodopa in Parkinson's diseas, 91
  - potentiation of phenyalanin in depression, 88
  - safeness, 90
  - slowing the progress of Alzheimer's disease, 92–94
  - slowing the progress of Parkinson's disease, 30, 89–91
  - transdermal application, 88
  - tyramine-inhibiting property, 28, 29
  - uptake inhibitory effect, 32
- 3-F-BPAP: lower activity derivative of BPAP, 47, 48
- mesencephalic enhancer substances, 25–39
- natural enhancer substances, 14, 25–27
- β-phenylethylamine (PEA), 25, 26, 49, 73–76, 119, 145
- (–)-PPAP: analogue of (–)-deprenyl, 30, 32, 35
- synthetic enhancer substances, 27–39, 145
- tryptamine, 25–27, 49, 74, 119, 145
enlightenment, 4, 133, 138, 140
ersatz, 98, 129

escape failure (EF), 45, 46, 56
essential changes during the life of mammals, 62–66
estrone, 70, 71
extinguishable conditioned reflex (ECR), 5, 13, 21, 51, 113, 118, 142

F

3-F-BPAP, *see* enhancer substances
fast-scan cyclic voltammetry (FSCV), 114
Fauve era, 126
fear, 113
fight or flight, 8, 116, 142
fluoxetine, 38
food deprivation, 15, 19
forgetting, 98–101
France, 135
free radicals, 63
French Revolution, 121–124

G

GABAergic pathways, 89
Germany, 2, 135, 136
Gestalt psychology, 1
ghetto, 136
Ginkgo biloba extract, 92
glass-cylinder-seeking drive, 2, 12, 13, 17, 51, 96, 141
glass-cylinder-seeking rats, 13, 19, 53, 54, 102, 103, 105–109, 111, 114
glutamatergic input to neostriatum, 89
goal-seeking behavior, 8, 14, 116
goals
- dispensable, 11
- indispensable (vital), 11

H

hallucinogens, 2
hate, 113, 135

hepatic encephalopathy, 74
hippocampal, hippocampal neurons, 27
Holocaust, 135, 136
Holy Scriptures, 140
home/school/society triad, 129, 131
homeostasis, 11
Homo sapiens, 3, 6, 7, 22, 54, 129, 136, 140, 141
HPLC – high pressure liquid chromatography, 32
human society, 3, 21, 124, 128, 138, 142, 143
hypertension, 28, 29, 74

I

imipramine, 28
impressionism, 125
individual differences in behavioral performances, 95–98
inextinguishable conditioned reflex (ICR), 5, 7, 21, 23, 50, 52, 53, 97, 99–101, 113–117, 125, 129–131, 133, 135, 136, 138, 142–144
innate drives, 2, 7, 11, 18, 139, 141, 144
insight, 5, 6
Internet, 134
intersignal reaction (IR), 45, 46, 56
iproniazid, 28
irrational brain activity, 143
isoniazid, 28
Israel, 135

J

Japan, 133
Jerusalem, 135
Jewish Holy Scriptures, 136
Jews, 135
Jordan, 135
joy, 113

K

Koran, 136

L

lazabemide, 38
learning, 49–56, 58–60, 62, 118, 119
– as a cortical enhancer regulation dependent, 56–60, 62
– modification of behavior through exerciese, training or practice, 49–55
levodopa, 30, 85, 91
Lewy bodies, 87
locus coeruleus, 15, 32
longevity, 63–66, 72
– developmental, 62–65
– postdevelopmental, 8, 9, 62–66, 131

M

mass-effect in humans, 115, 135
melanin granules in neurocytes of the substantia nigra, 86, 87
– area of one granule, 87
– density features, 87
– number, 87
– total area, 87
methamphetamine, 32, 36, 74
microglia, 42
migraine, 74
monoamine oxidase (MAO), 29, 74
– MAO "type A", 29
– MAO "type B", 29, 74, 75
monoamine oxidase (MAO) inhibitors, 2, 28
– MAO-A inhibiton, 29
– MAO-B inhibiton, 29–33
Mount of Olives, 135
Muslims, 136
myth of an *other world*, 123, 130
myth-based era of human society (myths-directed society), 130, 134, 137–141

Index of Subjects

N

narrative period of fine art, 126
natural death, 8, 20, 62–64, 78, 79, 81, 93, 145
Nazi Germany, 128
neostriatum, 89
neurodegenerative diseases, 89–94, 145
neuroglial cells, 42, 44
neuropsychopharmacology, 2
neurotrophic factor
- brain-derived neurotrophic factor [BDNF], 42, 43
- glial cell line-derived neurotrophic factor [GDNF], 42, 43
- mRNA expression of NGF, BDNF, and GDNF, 43
- nerve growth factor [NGF], 42, 43
nigral degeneration, 90
nigrostriatal dopaminergic neurons, 20, 31, 86
Nobel Prize, 23, 123, 136
nonvigilant resting state (sleeping), 8, 111, 116, 142
nootropic drug, 93
nucleus accumbens, 114

O

oligodendroglia, 42
orienting-searching reflex, 15, 18, 19, 66, 67
origins of science and art, 120, 144
oxiracetam, 93
oxygen radicals, 31

P

Palestinians, 136
"Panem et circenses", 122
Parkinson's, 71
Parkinson's disease, 9, 20, 29–31, 74, 75, 89–91, 145
- de novo disease, 29, 31, 91
- prevalence, 93
pentazocine, 49
peptidergic striatal interneurons, 89
pergolide, 38
phenothiazines, 2
β-phenylethylamine (PEA), *see* enhancer substances
phenylketonuria, 74
phosphatidylserine, 93
population growth, 3, 4
(−)-PPAP, *see* (−)-pdeprenyl
problem solving, 6
process of imagination, 5, 6
professional drives, 12
progesterone, 70
progressive decay of behavioral performances, 73–78
pyramidal neurons, 92

R

radio-ligand binding assay, 49
raphe, 15, 32
rational brain activity (order in the brain), 137–140, 142
rationally organized human society, 137–140
remembering, 98–101
reserpine, 81
Rett's syndrome, 74
reverberating circuits, 5

S

Sapere ande! (Dare to go independently!), 4, 129, 140
scavanger function, 63, 86
schizophrenia, 73
self, 5, 131
senses, 118
serotonergic neurons, 14, 20, 67–71
serotonin, 15, 25, 67–69

sexual activity in human males, 76, 77
sexual activity in male rats, 76–78, 84, 95, 96, 104, 106
– ejaculatory, 76–78, 84
– intromission, 76–78, 84
– mounting, 76–78, 84
sexual hormones dampen enhancer regulation, 20, 66–69
sexual maturity, 67–69, 131
shuttle box, 44–46, 56–60, 62
sigma agonists, 49
sigma receptors, 49
sleeping, see nonvigilant resting state
slegiline, see (−)-pdeprenyl
sorrow, 113
soul, 5
Spain, 135
striatum, 15, 32
substance P, 49
substantia nigra, 15, 32
supplementary drives, 12

T

talent, 51
Temple Mount, 135
terrorism
– March 11, 2004, Madrid, 137
– September 1, 2004, Beslan, 137
– September 11, 2001, New York, 137
– suicide killers, 23, 137
testosterone, 70, 71
tetrabenazine, 47, 48, 59, 60, 62, 89
TLS – technical life span, 65, 71, 76
α-tocopherol, 31, 32, 92

trace-amine receptors, 48, 49, 61, 74
traditional medicine, 23
transition from life to death, 64, 65
transition from uphill to downhill period of life, 64–71, 145
trial and error, 5, 18, 141
tricyclic antidepressants, 2
tuberculum olfactorium, 15, 32
TV image analyzer, 86

U

UN General Assembly resolutions, 135
UN Security Council resolutions, 135
uphill period of life, 8, 19, 131
uptake inhibitors, 2

V

vigilant resting state (leisure), 8, 116, 142
vis vitalis, 26, 66
vitamin E, see α-tocopherol

W

weaning, 19, 62–64, 67, 68
weapons of mass destruction, 137
What Are We?, 1, 140
Where Are We Going?, 1, 140
Where Do We Come From?, 1, 140
World War I, 135
World War II, 128, 133, 136

Y

youthful power, 8, 62, 67–69